D1580516

TEDBooks

How We'll Live on Mars

STEPHEN L. PETRANEK

TED Books
Simon & Schuster

London   New York   Toronto   Sydney   New Delhi

First published in Great Britain by Simon & Schuster UK Ltd, 2015
A CBS COMPANY

First TED Books hardcover edition July 2015

TED BOOKS and colophon are registered trademarks of TED
conferences, LLC.

For more information on licensing the TED talk that accompanies
this book, or other content partnerships with TED, please contact
TEDBooks@TED.com

1 3 5 7 9 10 8 6 4 2

Simon & Schuster UK Ltd
1st Floor
222 Gray's Inn Road
London WC1X 8HB

www.simonandschuster.co.uk

Simon & Schuster Australia, Sydney
Simon & Schuster India, New Delhi

A CIP catalogue record for this book is available from the British Library

Hardback ISBN: 978-1-47113-888-1
Ebook ISBN: 978-1-47113-889-8

Interior design by MGMT.design
Jacket design by Lewis Csizmazia

Printed and bound by CPI Group (UK) Ltd, Croydon, CR0 4YY

I want Americans to win the race for the kinds of discoveries that unleash new jobs . . . pushing out into the solar system not just to visit, but to stay. Last month, we launched a new spacecraft as part of a reenergized space program that will send American astronauts to Mars.

—President Barack Obama, State of the Union Address, January 20, 2015

CONTENTS

How We'll Live
on Mars

The Dream

A prediction:

In the year 2027, two sleek spacecraft dubbed *Raptor 1* and *Raptor 2* finally make it to Mars, slipping into orbit after a grueling 243-day voyage. As *Raptor 1* descends to the surface, an estimated 50 percent of all the people on Earth are watching the event, some on huge outdoor LCD screens. At this juncture of Earth's and Mars's orbits, it takes about twenty minutes for signals to reach the home planet, so people back on Earth are caught in a strange warp of time and space. As the ship descends to Martian soil, the four astronauts aboard could already be dead if something has gone wrong.

Nearly a decade of anticipation has come down to this moment: the spacecraft inches to the surface as the blast effect of braking rockets kicks up red dust. An Earth-bound audience waits eagerly as an announcer reminds them of a press conference that took place years earlier—a meeting that shocked the world and embarrassed NASA, which was still at least two years from testing its Mars spacecraft with humans aboard. On that day, the company behind this private effort to reach Mars revealed that it was about to build a series of huge rockets to transport people to Mars, and that within a decade it would launch one or two of them to effect the first manned landing on the Red Planet.

As *Raptor 1* settles into a massive crater near the Martian equator, the astronauts aboard are already thinking ahead. Time is precious. If all has gone well with the first landing, *Raptor 2* will follow within hours, carrying more explorers on board. First on the astronauts' punch list is the deployment of a base camp habitat, part of the enormous cargo the ships have carried. They must also inflate "buildings"—domed, pressurized tents made of exotic materials that will increase their living area and act as greenhouses in which to grow food.

Some environmental similarities exist between Earth and Mars. The Martian terrain looks a lot like certain parts of Earth—the dry valleys in Antarctica or the high deserts on Hawaiian volcanoes. Many other factors will prove to be extremely challenging. A day on Mars is only thirty-nine minutes and twenty-five seconds longer than a day on Earth, but a Martian year is far longer than one on Earth—687 days— making seasons twice as long. Mars's orbit is oval, meaning seasonal variations between winter and summer are more severe than those on Earth; in the southern hemisphere, summers are warmer and winters are colder. Ultimately these Martian settlers intend to establish two bases, one below the equator in the southern hemisphere for summers and one north of the equator for winters.

But now, within twenty-four hours, the first humans to walk on Mars must begin their most important task: finding water. They must determine if there is, as NASA landers and orbiters have predicted, enough water in the surface soil, called regolith, to support both their hydration needs as well as serve as a stock for making more of the oxygen they will consume. The astronauts purposely landed in a crater that a NASA orbiter spotted as having a smooth sheet of pure water ice. If that sheen isn't ice, they will need to find a second site nearby with a high percentage of ice in the regolith. If that sort of ice cannot be found nearby, the astronauts will deploy ground-penetrating radar to find underground water, and then drill.

Long before the next ships arrive (two years from this moment), these astronauts must build more permanent structures, possibly out of bricks they make from the regolith. Although today is sunny and relatively warm—about 50 degrees Fahrenheit—temperatures will plunge as darkness approaches, turning the environment into something akin to a bad night at the South Pole. Landing near the equator allows the astronauts to take advantage of milder temperatures that can reach 70 degrees Fahrenheit on a summer day. But at night, the temperature easily reaches minus 100 degrees, and structures will be needed to insulate the astronauts from the cold as well as to protect them from solar rays that are almost unrestricted by the thin atmosphere.

In the event that everything goes wrong—they cannot find a good water supply, the radiation effects are more severe than predicted, or one of the ships is badly damaged on landing, they will hunker down and await a decent launch window for the long trip back to Earth. Otherwise, they are here to stay.

These first explorers, alone on a seemingly lifeless planet as much as 250 million miles away from home, have everything in common with the trailblazers who preceded them—the great explorers throughout history who scaled mountains and sailed oceans to create new lives. And yet, despite their commonality with explorers of years past, these spacefaring pioneers are in every way more important than any explorers who have ever preceded them. Their presence on Mars represents the greatest achievement of human intelligence.

Anyone who watched Neil Armstrong set foot on the moon in 1969 can tell you that, for a moment, the Earth stood still. The wonder and awe of that achievement was so incomprehensible that some people still believe it was staged on a Hollywood set. When astronauts stepped onto the moon, people started saying, "If we can get to the moon, we can do anything." They meant that we could do anything on or near Earth. Getting to Mars will have an entirely different meaning: If we can get to Mars, we can go anywhere.

The achievement will make dreamy science fiction like *Star Wars* and *Star Trek* begin to look real. It will make the moons of Saturn and Jupiter seem like reasonable places to explore. It will, for better or worse, create a wave of fortune seekers to rival those of the California gold rush. Most important, it will expand our vision as far from the bounds of Earth's gravity as we can imagine. When the first humans set foot on Mars, the moment will be more significant in terms of technology, philosophy, history, and exploration than any that have come before it, all because we will no longer be a one-planet species.

These explorers are the beginning of an ambitious plan, not just to visit Mars and establish a settlement but to reengineer, or terraform, the entire planet—to make its thin atmosphere of carbon dioxide rich enough in oxygen for humans to breathe, to raise its temperature from an average of -81 degrees Fahrenheit to a more tolerable 20 degrees, to fill its dry stream beds and empty lakes with water again, and to plant foliage that can flourish in its temperate zone on a diet rich in CO_2. These astronauts will set in motion a process that might not be complete for a thousand years but will result in a second home for humans, an outpost on the farthest frontier. Like many frontier outposts before it, this one will eventually rival the home planet in resources, standard of living, and desirability.

These pathfinders have embarked on a journey that has implications far into the future. Their greater mission is to establish a spacefaring society that maintains a system of spaceports for rockets, allowing easy liftoff from a planet with low gravity. From there, humans can travel to the outer reaches of the solar system.

When these rockets land on Mars in the near future, it will be far more than a great moment for exploration. It will be nothing less than an insurance policy for humanity. There are real threats to the continuation of the human race on Earth, including our failure to save the home planet from ecological destruction and the possibility of nuclear war. Collision with a single asteroid could eliminate most life, and eventually our own sun will enlarge and destroy Earth. Long before that happens, we must become a spacefaring species, capable of living not only on another planet but ultimately in other solar systems. The first humans who emigrate to Mars are our best hope for the survival of our species. Their tiny base will grow into a settlement, and perhaps even a new species that will expand rapidly. The company which built the rocket that brought them there is building hundreds more rockets. The intention is to create a viable population of 50,000 within a few decades. They can preserve the collective wisdom and achievement of humanity even if those of us back on Earth are annihilated.

The truth is that it has been possible to reach Mars for at least thirty years. Within a decade or so of the *Apollo 11* mission that landed the first humans on Earth's moon, we could have landed humans on the Red Planet. Almost every technology required has long been available. We simply have not chosen to pursue the opportunity.

The backstory of that mistake is worth understanding—how a single decision by one US president stunted space travel for decades, how we might have inspired two generations of earthlings with humankind's ability to deliver on almost anything our brains can imagine. Nearly five decades ago we had the ability to extend ourselves into the solar system and beyond.

Now private rocketry has opened a new window to the stars. Perhaps the need to explore is built into our DNA; *homo sapiens* began venturing out of Africa about 60,000 years ago, pushing beyond the horizon until they populated the entire globe. Exploration may be connected to human survival. But it has also led to colonization of lands already occupied, the devastation of cultures, and the plundering of resources.

The settlement of Mars is about to happen far sooner than most people realize, and in a nonregulated way. Most of this book is an examination of the astonishing fact that we have the capabilities to build on Mars. But this book is also a wakeup call. The potential is enormous, but the pitfalls are numerous. The time to think is now.

1 *Das Marsprojekt*

When Robert Goddard launched the first liquid-fueled rocket to a grand altitude of forty-one feet in 1926, could he have possibly imagined we'd be touching down on Mars 101 years later? Yet the path is surprisingly straightforward. You can draw a line from those first astronauts that will land on Mars in 2027 straight back to one person: a former SS officer in World War II named Wernher von Braun. As his designs, built on Goddard's inventions, rained ruin on London, von Braun's genius for rocketry became all too obvious. He gave Adolf Hitler weapons of terror that shocked the world. Yet by 1948, only four years after von Braun's sophisticated V-2 rocket was launched across the North Sea, the thirty-six-year-old engineer found himself living at Fort Bliss, Texas, with a team of German rocket-scientist colleagues, all of them prisoners of peace.

The US Army team that spirited von Braun and his engineers out of Germany reportedly allowed them to leave the base only with an escort, and so von Braun and his men spent their days uploading their extraordinary expertise into the minds of Americans who were trying to build a ballistic missile. But often they had little to do. So the former head of the world's most advanced rocket program decided to write a book about his favorite subject—space exploration. The book wasn't published until 1952, and then only in German, as *Das Marsprojekt*. In 1953 the University of Illinois Press published it in English as *The Mars Project*. To this day, the ninety-one-page brief remains the most influential manual on space travel ever written. It has never become obsolete, and much of it still serves as a guide to getting humans to Mars.

The vision von Braun set out in his book was massive—it involved seventy men flying on a fleet of ten spacecraft, three of which were cargo ships that would not return.

"I believe it is time to explode once and for all the theory of the solitary space rocket and its little band of bold interplanetary adventurers," von Braun wrote. "No such lonesome, extra-orbital thermos bottle will ever escape earth's gravity and drift toward Mars."

The plan was to build the spacecraft at a space station in Earth's orbit. The equipment and materials would be launched there by forty-six fully reusable three-stage rockets, the first two stages of which would parachute back to terra firma, while the third would fly home on wings. This vision—*created in 1948*, when von Braun made most of his calculations—predicted the US space shuttle, as well as Space Exploration Technologies Corporation's current efforts to build a reusable orbital rocket that can be refueled and relaunched within twenty-four hours. In 1953, von Braun estimated that nine-hundred-fifty ferry flights would be necessary to build and fuel the ten spaceships.

The German engineer's version of a trip to Mars called for the use of Hohmann transfer orbits—a fuel-saving method by which a spaceship in a circular orbit of Earth can briefly turn on its engines—called a burn—to escalate into an elliptical orbit that will intersect with Mars's orbit around the sun. The ship can then coast without burning fuel until it gets close to Mars. Once near Mars, a second burn will decelerate it into an orbit around the Red Planet. It's a bit like Tarzan swinging on a long vine to get between two trees that are far apart, and then picking up a short vine to get to a specific branch. The maneuver requires precise timing of the alignment of Mars and Earth in their orbits.

A good window for a Mars launch occurs about every twenty-five months, but the coasting economy of the Hohmann transfer orbit comes with a price—the trip takes about eight months each way. Every fifteen years or so, the orbits of Mars and Earth align in such a way that the distance between the planets (and thus the travel time) is greatly diminished. Other theories of how to get to Mars on the least amount of rocket fuel now abound, including simply burning a lot of fuel to make the journey in a straighter line, shorter than the curve of a Hohmann transfer orbit. Unproven and mostly theoretical propulsion systems such as nuclear fusion and nuclear electric propulsion could shorten the trip to fewer than ninety days each way.

Had von Braun's mission launched, astronauts would have been forced to spend approximately four hundred days exploring Mars, until the Earth slipped into the right place for the journey home on another Hohmann transfer orbit.

Das Marsprojekt was written long before scientists knew about the shielding effect of Earth's protective Van Allen radiation belt (highly charged particles that do not exist near Mars), the effects of prolonged weightlessness, the severity of solar radiation (von Braun had made calculations for cosmic radiation), the actual geography of the Red Planet, and anything but the most rudimentary estimates of the density of the atmosphere on Mars. Von Braun's book preceded by a decade the first orbiter in space—*Sputnik*, launched in 1957. In the manual, he admitted that he had not calculated the dangers of meteors, but he had worried about weightlessness and had drawn up plans for connecting his flotilla of spacecraft together with cables and then spinning them around one another like so many yo-yos to produce artificial gravity.

When NASA's *Mariner 4* probe flew past Mars in 1965, it relayed back two shocking facts: the Martian atmosphere was much thinner than scientists had assumed—making it practically nonexistent—and there was little chance of anything being alive. Von Braun, like many people on Earth in the 1960s, had toyed with the idea of an alien species living on Mars in underground gardens. He even wrote about such a civilization in a romantic and corny novelized version of *Das Marsprojekt* in 1949. To get from Mars orbit to the surface, von Braun had designed small space planes. Although they wouldn't have been flyable in such a thin atmosphere, he had foreseen such unexpected difficulties. He provided numerous backup strategies for the mission, and the planes were designed so that their wings could be jettisoned.

Von Braun also imagined there would be psychological drawbacks of long-term travel among humans confined in a small space for months or years. He created shuttle craft to move supplies from ship to ship during the voyage to Mars as well as to transport astronauts to and from different ships. Later calculations based on his plans allowed for 26,500 pounds of oxygen per person, 17,500 pounds of food, and 29,000 pounds of potable water per passenger vessel. Each ship would be able to recycle utility water and water vapor in the air.

In the technical appendix to the book, one statistic about spaceflight stands out—the extraordinary amount of fuel that is necessary to escape Earth's strong gravity. Each of the ten ships in von Braun's fleet would weigh about eight million pounds, a bit more than seven million pounds of which would be rocket fuel. Each ship would weigh just over 1 percent of its original weight when it returned to Earth from Mars.

Das Marsprojekt was a work of extraordinary foresight and sheer engineering genius. Unfortunately, both von Braun and Robert Goddard were so far ahead of their time that they suffered a great deal of bad publicity, criticism, and abuse from authority figures who simply didn't get it. When Goddard said that a rocket could reach the moon, the news made the front page of the *New York Times*, but on its editorial page, the paper mocked his vision. (Nearly five decades later, and a day after *Apollo 11* took off for the moon, the *Times* printed a correction.)

In the early 1950s, when von Braun proposed a serious plan for going to Mars, his vision must have been experienced as absurd—even by scientists and engineers. Hundreds of rockets sent into low Earth orbit to build ten giant interplanetary spacecraft filled with tens of millions of pounds of fuel and oxygen and food? Really?

But it certainly captivated the American public. In 1954

Collier's magazine published an eight-part series on space travel that included von Braun's description of exactly how we could get to Mars.

Plenty of dreamers had given serious thought to interplanetary space travel before von Braun, but no one had come up with designs and precise calculations. His proposal was complete with trajectories, equations, technical drawings, and calculations. He even worked out a date for the launch, in 1965. His mind took a no-nonsense approach—the only difference between imagining a trip to Mars and actually going there was commitment.

To grasp the significance of *Das Marsprojekt*, one might reflect on Carl Sagan's 1985 novel *Contact*. In that book of fantasy, an unknown culture somewhere in space and time beams a technical manual to humans that describes in detail how to build a spacecraft that can reach their alien world. To most earthlings in the early 1950s, von Braun would have been that alien culture, gifting us with a technical manual for exploring the universe. The difference is that *Das Marsprojekt* wasn't fanciful.

By the late 1960s, von Braun was widely revered for masterminding the Saturn V rocket—the rocket that got the *Apollo* astronauts to the moon. As his reputation grew, von Braun took to thundering through the halls of NASA and Congress, insisting that the United States shoot for the Red Planet next. This time his proposal called for sending two spacecraft, powered by nuclear engines. He told anyone who would listen that the mission could be launched in the 1980s.

Unlike the proposals von Braun had made in the past, this

one reached the desk of President Richard Nixon. There had to be a plan for what would come after *Apollo*. But von Braun's Mars plan lost out to the space shuttle program—in part because military and intelligence agencies thought the space shuttle could be extremely useful for launching and repairing spy satellites. Although everything NASA does is supposed to be completely transparent and public, in the decade from 1982 to 1992, the agency launched eleven classified shuttle missions. Much of the shuttle's design was driven by military and intelligence agency requirements. Nixon also decided to kill von Braun's Saturn V rocket, the largest and best heavy-lift rocket ever conceived. Without that ship to sail on, interplanetary travel was doomed. Had the United States committed to Mars instead of the space shuttle, we likely would have a permanent base there now. Von Braun, seeing that he and NASA were headed in different directions, retired from the agency in 1972.

In the sixty-two years since *Das Marsprojekt* was first published, Mars has become little more to us than a series of photos taken by an automobile-sized rover called *Curiosity* that has been exploring its surface since landing in 2012. Had NASA and Nixon taken von Braun's ideas more seriously, *Curiosity* might be roaming Mars with astronauts at the wheel.

The space shuttle threw the US space program into a long, slow decline and sent the agency, if not the American people, into a void that lacked passion and vision. The inspiration of reaching for the stars was replaced by space walks no one watched. NASA focused on a dated rocket plane design that did little to advance space travel, fell out of the sky far too often, and had nowhere to go until a space station was built for it to dock with. Its final, weak justification was to ferry astronauts and cargo to the International Space Station (ISS), which itself is a pretty useless hunk of technology.

Sir Martin Rees, the much-admired Astronomer Royal of Britain, has few good words to say about the International Space Station: "No one would regard the science on the Space Station as being able to justify more than a fraction of its overall cost. Its main [purpose] was to keep the manned space programme alive, and to learn how humans can live and work in space. And here again the most positive development in this area has been the advent of private companies which can develop technology and rockets more cheaply than NASA and its traditional contractors have done."

NASA's years of slip-sliding away—and the fact that it allowed contractors to work on a cost-plus basis—created a huge hole for entrepreneurs to walk through. The 135 shuttle missions ended up costing an average of more than $1 billion each. Somebody ought to be able to do what NASA did better, faster, and cheaper. Those somebodies have arrived. And they've made the fantasy of going to Mars a reality.

2 The Great Private Space Race

Getting into space has always seemed the province of governments that can afford the high cost of entry. The Boeing and Lockheed Martin space businesses, for example, largely happened because NASA and the US military were willing to write cost-plus contracts. Then, about thirty years ago, three men who met at Harvard Business School decided they could make money building rockets and launching satellites. Their start-up, Orbital Sciences Corporation, designed a unique three-stage rocket with a wing and slung it underneath the wing of a large passenger jet. The rocket was dubbed Pegasus. The jet flew the rocket to forty thousand feet, providing a cheap boost into orbit. Since then, Pegasus has flown forty-two times and built an extraordinary success record, with only three complete failures. Orbital Sciences has carved a successful niche for itself, designing and building rockets and satellites. It has launched and built hundreds of satellites and probes for telecoms and governments and NASA, some of them with reconfigured intercontinental ballistic missiles (ICBMs). In recent years, thanks to NASA contracts, advice, and encouragement, Orbital Sciences built a new rocket called Antares and a spacecraft called *Cygnus* that has successfully delivered supplies to the International Space Station at a small fraction of the cost of using the space shuttle. The company makes a profit, has merged with another spacecraft builder, and is listed on the New York Stock Exchange as Orbital ATK.

As Orbital Sciences was building its business, an aerospace engineer at Martin Marietta Materials named Robert Zubrin got antsy about why we weren't headed to Mars. Zubrin thought long and deeply about what would be necessary to make Mars habitable, and his calculations have added a level of sophistication to the discussion. He foresaw, as von Braun did, that everything we needed was at hand. His proposal—called Mars Direct—outlined plans for an inexpensive and simple manned mission to Mars. NASA liked it, but when the agency kept delaying action, Zubrin wrote a comprehensive book called *The Case for Mars* and formed the Mars Society in 1998 to help promote his idea.

More recently, Dutchmen Bas Lansdorp and Arno Wielders formed the nonprofit Mars One to launch one-way trips to the Red Planet, which they say are scheduled to land in 2025 (after previously landing cargo craft, habitats, and rovers). It plans to pay for the venture by selling broadcast rights. However, the group not only doesn't have a rocket or spacecraft that will get it there, it has only recently signed a contract with Lockheed Martin to study the feasibility of creating such things.

Then there is Dennis Tito, the first private citizen to buy his way into space by paying the Russians a reported $20 million. His nonprofit organization, Inspiration Mars, optimistically plans to send a small spacecraft—perhaps the *Crew Dragon* spacecraft under development by SpaceX for manned flights to the International Space Station—to Mars with a married couple aboard in 2021. The mission is a flyby, so the couple will be stuck in a tiny capsule for nearly a year and a half straight—which is precisely why Inspiration Mars proposes to send a husband and wife. To overcome deprivation and loneliness, Tito says, "you're going to need someone you can hug."

Inspiration Mars is aiming for a 2018 launch because a once-every-fifteen-years positioning of Mars and Earth in their respective orbits would permit a 501-day round-trip flyby using a single trajectory burn. The rest of the trip would be a coast to Mars, a slingshot around the planet, and then a coast back to Earth. No useable rocket yet exists that could perform the feat. (NASA's Space Launch System, due in 2018, could lift a spaceship from Earth to Mars, but is unlikely to be loaned to the effort.) Tito says the backup plan is to launch in 2021 and slingshot around Venus into a trajectory for a Mars flyby.

Amazon's Jeff Bezos, Google cofounder Larry Page, Microsoft cofounder Paul Allen, and entrepreneur and explorer Sir Richard Branson are also investing millions to get into the new private space race in one form or another. Most of the efforts so far are as chaotic as the Wild West, but this time the frontier is space. And although there is no shortage of private projects intending to send people to Mars, only one company can currently make a realistic promise to deliver human bodies to the Red Planet before NASA finally gets around to it.

• • •

In the same way we can draw a line from Wernher von Braun straight to *Apollo 11*, when a spaceship carrying astronauts lands on Mars in 2027, we may well be able to draw a line straight to Elon Musk—because that Mars lander will most likely have the SpaceX logo on it.

Musk is arguably the most visionary entrepreneur of our time. Seven years after he quit a PhD program in applied physics at Stanford University, he sold his share of PayPal and Zip2, companies he cofounded, giving him a reported net worth of $324 million. He rolled his money into Space Exploration Technologies Corporation (SpaceX), a company he founded in 2002, then went on to cofound Tesla Motors, which is poised to revolutionize the automobile world. He is a devout environmentalist and proponent of solar energy—his Teslas can literally be driven on sunlight. In 2013, Musk proposed a unique high-speed transportation system in a vacuum tube called Hyperloop, which he put into the public domain. A Hyperloop tube running between Los Angeles and San Francisco could reduce travel time to thirty minutes.

Musk formed SpaceX just when it seemed as if NASA was slipping into irrelevance. Like von Braun, he is a transplant, in this case from South Africa and Canada. Musk, like von Braun, is a perfectionist who is convinced of his vision and determined to achieve it. And as with von Braun, no one seems to understand how serious Musk is when he says we must get to Mars. Against all advice and all odds, he has managed to do the impossible: find enough capital to finance Space Exploration Technologies and to keep it afloat and moving forward even when its first three rockets blew up. Along the way, he has raised a truly revolutionary question: Who needs NASA to get to Mars?

There is only one reason Musk started his private rocket company: "The reason SpaceX was created was to accelerate development of rocket technology, all for the goal of establishing a self-sustaining, permanent base on Mars," he said in May 2014. Let us pause, for a moment, to look twice at the name of Musk's company—Space Exploration Technologies. Note the word *Exploration*. Like von Braun before him, Musk is in love with the idea that humans should become a spacefaring society. He is keenly aware that Earth will not be habitable forever. Musk seems frustrated by our denial about what we are doing to our habitat, and is ever cognizant of a simple fact: humans will become extinct if we do not reach beyond Earth.

Elon Musk's appearance as a rocket man came none too soon. The technology had advanced very little from 1969, when Neil Armstrong placed his boot on the moon, to 2002, when Musk began SpaceX. In fact, according to Musk, space travel since the Apollo program not only hasn't moved forward much, it has gone "backward." He says, "Once we could go to the moon, and now we can't. That's not forward, or even sideways. The United States can't even send people into orbit right now."

In 1966 NASA's budget was more than 4 percent of the total federal budget. Today it is about half of 1 percent. Since Musk appeared on the space scene, we have been moving at the speed of light toward the technical solutions needed to get humans to the Red Planet in just over a decade and to potentially keep settlers there for millennia to come. No one can point a finger at the exact moment when NASA finally woke up and smelled the red dust, but when SpaceX's first *Dragon* capsule successfully reached the International Space Station in May of 2012, it immediately became obvious that a private company could probably do anything NASA could—and perhaps do it better.

3 Rockets Are Tricky

Recently, after one of his rockets exploded just above its launch pad, Elon Musk wryly tweeted: "Rockets are tricky." He's right: close to two-thirds of all the attempts to get probes to Mars have failed.

A casual observer might well wonder why humans have had so much trouble getting to Mars when getting to the moon more than fifty years ago seemed relatively easy. Mostly, it's a matter of distances. The scale changes are phenomenal. The moon floats between 225,000 and 250,000 miles from Earth, depending on the lunar cycle. Mars can be up to a thousand times farther away. In 2003, Mars and Earth were closer than they had been in almost sixty thousand years—only about 34 million miles apart. But because Earth's orbit around the sun takes 365 days and Mars's takes 687 Earth days, the two planets can get out of sync and wind up very far apart, with each on a different side of the sun. When they are far apart, they are *really* far apart—about 250 million miles. Mars thus varies between being 140 and 1,000 times farther away from Earth than the moon.

Put another way, humans can make a round-trip to the moon and back in six days. (We could have gotten there in one day with the boost the Saturn V rocket offered, but we would have been going so fast when we arrived that we would simply have shot by instead of being captured by the moon's weak gravity.)

Using the Hohmann transfer orbits suggested by von Braun in *Das Marsprojekt*, even if we went much faster than the speed at which the Apollo astronauts went to the moon, we would still have to fly about a thousand times farther than the distance to the moon to end up at Mars. That's because we simply can't carry enough fuel to blast ahead in a straight line. Without unlimited cheap energy, we will always be in orbit around *something* in this solar system, so all our trajectories will be curved. There are no foreseeable shortcuts in the next twenty years that could get us to Mars in much less than 250 days each way, although SpaceX is designing more powerful and more efficient rocket engines that could shorten the trip substantially.

Even the early, more straightforward missions to Mars— missions that merely attempted to fly by the planet—regularly met with disaster. The far more difficult Mars orbiter missions, and especially the lander missions, made something of a mockery of our grasp on space technology.

The Soviets seemed to get the worst of the early Martian calamities. The first Earth object ever to reach the surface of Mars was a Soviet lander called *Mars 2*. It crash-landed in November of 1971, and was a follow-up project to *Kosmos 419*, which never got out of orbit around the Earth, much less headed to Mars. The next month, *Mars 3* actually made a successful landing but stopped sending signals after twenty seconds. *Mars 4*'s guidance system failed, and it whizzed by the planet completely. *Mars 5* was the most successful Soviet probe. It was inserted into an elliptical orbit in February 1974, and returned about sixty photos during twenty-two orbits, then failed. *Mars 6* reached

the planet in March of 1974 and launched a lander that crashed on the surface. It transmitted atmospheric data for about four minutes before it went silent, but the data was largely incomprehensible because of a computer chip failure. *Mars 7* also entered orbit in March 1974 but launched its lander four hours too early and missed the planet. There were a handful of other earlier Mars missions launched by the Soviets that failed, as well as later failed missions. In 1996 the Russian Space Agency launched an orbiter/lander called *Mars 96* that didn't escape Earth's gravity and broke up over the Pacific Ocean. Since then, the Russians have seemed less than eager to challenge their jinx.

A huge hindrance to successfully landing a probe on Mars is that it takes communications a long time to arrive from Earth. When Earth and Mars are farthest apart, it takes a radio signal twenty-one minutes to get from Earth to Mars, and then another twenty-one minutes for a return signal to get back to Earth. Unmanned spacecraft must therefore use artificial intelligence software to make decisions in emergencies, because there's no time to call home for help.

But all the bad history of early lander mission failure slipped into the darker reaches of our consciousness after NASA scored big by successfully landing the *Spirit* and *Opportunity* rovers on Mars. More recently, the success of the *Curiosity* rover has stolen our attention. *Opportunity* is still actively exploring Mars after more than a decade. *Curiosity* finished a Martian year's (just under two Earth years) worth of exploration in 2014, and is just getting started on its longer mission. Nevertheless, the distances these rovers have covered is not impressive. *Opportunity* has traveled only about twenty-six miles since 2004, and *Curiosity* has gone a bit more than six miles in nearly three years.

Despite the failures of the past, NASA's success with *Curiosity* proves that relatively large payloads can be delivered to the surface of Mars, making not only manned flights more realistic but also the idea of cargo and resupply flights. Changing the equation from large payloads like *Curiosity* to human cargo is mostly just a step up in scale, frequency of cargo launches, and oxygen. SpaceX is refining a *Dragon* spacecraft with the ability to carry seven astronauts that it expects to fly to the International Space Station as early as 2016, although Musk recently said that "2017 is probably a realistic expectation of when we'll send a human into space for the first time." He has joked that a stowaway astronaut on its current *Dragon* vehicle that resupplied the International Space Station would survive the flight because part of the craft is pressurized; it was designed from the start to be converted to carry astronauts instead of cargo.

Currently, the Russian *Soyuz* spacecraft is the only vehicle that can get astronauts to the space station and back in the absence of the space shuttle. It dates to 1966 and, along with the Soyuz rocket that carries it into space, has proven to be the most reliable space vehicle in history. As made famous in the movie *Gravity*, at least one *Soyuz* spacecraft is attached to the International Space Station at all times for use as an emergency escape vehicle. The Russians charge more than $50 million to fly an astronaut to the space station. SpaceX wants that business.

Late in 2014, NASA launched its new *Orion* spacecraft on a Delta IV rocket to an orbit of about 3,600 miles above Earth. *Orion* is designed to carry up to six astronauts on missions to the International Space Station and four astronauts to the moon and beyond. A more powerful rocket specifically designed for *Orion* should come online in 2018. The spacecraft looks a lot like an *Apollo* capsule and appears to be not much more than a second-generation moon-shot vehicle. Elon Musk says he is not aware of any facets of the *Orion* design that make it anything more than a larger *Apollo* spacecraft. Experts have defended it by saying that a proven design will lower risks.

Orion was designed to explore the moon and rendezvous with an asteroid in the 2020s. NASA has been remarkably circumspect about plans for a manned mission to Mars until very recently. It now vaguely forecasts Mars as the ultimate use of *Orion*, but will not commit to much of a timetable except to say that a Mars mission might happen in the 2030s. NASA has always stuck by the position that we should build a base on the moon first to learn more before attempting to do the same on Mars. At the rate *Orion* is being developed, Musk, and perhaps other private rocketeers, are positioned to get to Mars long before NASA does.

Nevertheless, the development of two different spacecraft that could get humans to Mars—SpaceX's *Dragon* capsules and NASA's *Orion*—has changed the basic question that's been floating around since *Das Marsprojekt* was written: Can we get to Mars? The answer is yes. The new question: Can we live on Mars? The answer to that is yes, too, but as Elon Musk might say, it's tricky.

4 Big Questions

At this point in time—less than two decades away from landing on Mars—there are still many skeptics. People in the space business tend to say that we should go to the moon first to set up a practice base there. Or they say the difficulties of making Mars a livable place for humans are too much to grasp. The fact is, the prospect of landing on Mars comes with a lot of challenges.

So let's pause briefly to review a few of the most common questions that arise.

> **Can a small group of people travel together in extremely confined spaces under significant stress for nine months without killing one another**? Let's answer this with another question: Didn't life in diesel/electric submarines in World War II answer this question? Besides, knowledge of human psychology has advanced to the point that picking the right people for a Mars mission is no longer a challenge. We are very good at identifying the right people to become commercial pilots, Navy SEALs, and others who occupy critical positions where stress, judgment, and intelligence meet. Angelo Vermeulen, a space systems researcher who led a team of Earth-bound astronauts through a four-month simulation of living on Mars on the island of Hawaii, says: "It all boils down to crew selection. You need

to match skills as well as psychological compatibility. You can quickly see if problems will develop by simply putting people together for a week and giving them something challenging to do. If there are going to be problems, you'll usually see them. There's never a guarantee issues won't arise over a long period of time. But you should start with a crew that explicitly likes to work together and is resilient."

Will anyone be willing to pay the cost—estimated to be $5 billion to get there and $30 billion to establish a small base? Elon Musk has answered that question by putting his money where his mouth is. He has declared that SpaceX will not go public until its Mars rocket is flying. That rocket will be far larger than even SpaceX's next generation rocket, the Falcon Heavy, scheduled to fly in late 2015 or early 2016 with twenty-seven engines and a first-stage thrust three times greater than SpaceX's current rocket, the Falcon 9. In other words, he will not subject his company to influence from stockholders who desire profits until he's certain he can get to Mars. "The first mission will be very expensive," Musk admits, but he expects that future missions will be financed primarily by the people who are going. As von Braun said, a trip to Mars costs no more than "a minute fraction of our yearly national defense budget."

Can enough safety factors be built into the mission to make success 95 percent probable? Ask film director and explorer James Cameron, who recently set the world manned-submarine diving record in the Mariana Trench. He points out that if designers of risky machinery carefully tackle all the known and obvious problems, the inevitable unexpected problems will likely also be surmountable.

Will astronauts' bodies fall apart in a prolonged zero-gravity environment? This is still a significant challenge, but von Braun's suggestion of tethering multiple spacecraft together and spinning them to create artificial gravity could be viable on a Mars journey. A spacecraft could be designed in the shape of a wheel to spin and create gravity. But it's worth remembering that the trip to Mars is only about two months longer than the typical amount of time an astronaut spends on the International Space Station. During 2015 and 2016 Captain Scott Kelly from the United States and Russian cosmonaut Mikhail Kornienko will spend an entire year aboard the ISS, and we'll learn more about long-term effects on the human body from their mission. Overall, gravity on Mars is slightly more than a third of what it is on Earth, but scientists speculate that may be just enough for humans to survive. Furthermore, recent research indicates that many species could evolve to new environments far more quickly than once thought. A Martian community might adapt to low gravity within a few dozen generations.

What if astronauts get ill? Explorers who climb mountains and sail around the world learned long ago that it's a good idea to bring along someone trained in emergency medicine. But solo long-distance sailors who circumnavigate the Earth have shown that the vast majority of problems can easily be addressed by the supplies in a well-stocked medical kit and proper training. Space travel, however, is not ocean sailing, and some explorers may very well fall ill and die.

What about radiation? This is still a big bugaboo. The worst solar radiation comes from solar flares and the related coronal mass ejections of radiation from our sun. We have no technologies that can eliminate solar and cosmic radiation, but we can design emergency spaces in interplanetary vehicles that are specially shielded for events like solar flares. And there will be adequate warning to seek a protective habitat until the burst has passed. Elon Musk has proposed a spacecraft insulated by water. There are other strategies to deflect or absorb radiation, but inevitably, astronauts going to Mars will experience far higher doses of radiation than would ever be allowed on Earth. NASA officials are evaluating an increase in the limits of radiation astronauts are allowed to experience as a way to get them to Mars under operating guidelines. Once on Mars, where the atmosphere is of limited help and there is no magnetosphere or Van Allen belt to block radiation, people will have to spend most of their time in shielded environments or underground.

As spacecraft are seriously developed for a Mars landing, other significant questions may arise, and new answers will have to be found.

5 The Economics of Mars

If no one can get to Mars economically, no one will end up living there. It's worth noting that Elon Musk sees the entire viability of a Mars settlement resting on basic cost issues rather than the many environmental impediments, such as having no air to breathe, dangerous radiation, and whether water is accessible.

In late 2012, Musk gave a talk to the Royal Aeronautical Society in London focused on rocket technology—specifically, how the reusability of rockets, as proposed by von Braun in 1952, would dramatically change the economics of space travel, if not be the determining factor in whether people could live on Mars.

Noting that the cost of launching a Falcon 9 rocket is about $60 million, and that only 0.3 percent of the cost is rocket fuel, Musk said: "So, if we could use the same Falcon 9 rocket a thousand times, then the capital costs would go from being $60 million per flight to $60,000 per flight. Obviously, that's a humongous difference." A Falcon 9 rocket is not large enough to get even one voyager to Mars, but Musk was pointing out the incredible cost savings represented by rocket reusability, a factor that would be multiplied many times over when it comes to the type of huge rocket needed to establish a self-sustaining civilization on Mars.

If reusability isn't achieved, Musk said, "I just don't think we'll be able to afford it, because it's a difference between something costing a half a percentage each year of GDP and all the GDP." Then he added: "I think most people would agree, even if they don't intend to go themselves, that if we're spending something between a quarter to a half a percent of GDP on establishing a self-sustaining civilization on another planet [it] is probably worth doing. It's sort of a life insurance policy for life, collectively, and that seems like a reasonable insurance premium, and plus it would be a fun adventure to watch even if you don't participate. Just as, when people went to the moon, only a few people actually went to the moon, but in a sense, we all went there vicariously. I think most people would say that was a good thing. When people look back and say, what were the good things that occurred in the twentieth century, that would have to be right near the top of the list. So I think there's value, even if someone doesn't go themselves."

Answering questions after the lecture, Musk sometimes spoke as if he were the CEO of a successful airline company instead of a development-stage rocket company. Assuming that enough people might make the trip if he can make it affordable, he guessed that SpaceX could make money selling one-way tickets for $500,000 each. More recently he said that tickets would "hopefully be less than five hundred thousand dollars each, but something like that."

Musk imagines a typical Mars emigrant being someone in his or her forties who owns a middle-class home worth $500,000. Perhaps she hates her job and decides to sell everything to buy a one-way ticket to Mars from SpaceX, with enough money left over to finance a small business.

During the question-and-answer period in London, Musk said:

> There is definitely some amount of money that has to be spent establishing a base on Mars. Basically, getting the fundamentals in place. Call it the activation costs of a Mars base. That was true also of the English colonies. They really took a significant expense to get things started. You really didn't want to be part of Jamestown. It was not good. It took quite a bit of effort to get the basics established before the subsequent economics made sense. So there is that investment, and we'll need to gather the money to do that, but then once there are regular flights, that's right when you need to get the cost down into the half-a-million-dollar range for somebody to move to Mars, because then I think there would be enough people that would buy that—they'd just sell their stuff on Earth and move to Mars—to have it be a reasonable business case. It doesn't need to be many people—there's seven billion people on Earth, probably reach about eight billion by the end of the century—and the world on the whole is getting richer, so I think if only even one in ten thousand people decide that they want to go, that'll be enough; even one in one hundred thousand.

The figure at the far end of Musk's estimate—one in one hundred thousand people who choose to be pioneers—could mean a Mars settlement approaching eighty thousand people, about the size of a small city on Earth. That may seem wildly optimistic, but in answering a question from the audience, Musk said: "Forecasts are always tricky. If you asked somebody at the dawn of air flight, what are your market forecasts? I mean, they're going to be wildly wrong. Probably on the low side. Even probably the most optimistic people at the beginning of aviation would seem like pessimists today."

In fact, Musk imagines far more than eighty thousand people in a city on Mars. He envisions eighty thousand people going in a single trip. "We're not designing a system to send a handful of people," Musk told me in an interview. "We're designing a Mars colonial transport system—something that, once brought to fruition, will be capable of establishing a self-sustaining colony on Mars. It's a very big system. Version one of that system we would aim to complete before 2030. From 2030 to 2050 we'd have ten orbital synchronization events . . . which means that maybe in the course of twenty years there would be forty thousand or fifty thousand people there."

Musk says his proposed Mars Colonizer will have only two rocket stages: "There's the booster section, just to get out of Earth's gravity, and then there's the spacecraft, which would be an integrated upper stage and spaceship. With Falcon 9, the upper stage and the spacecraft are separate, but in the case of the Mars Colonizer they will be integrated. The booster will get the rocket halfway to Earth orbit, and then the upper stage will get it the rest of the way. Then you'd have a tanker craft [in Earth orbit] that would replenish the propellant."

A large mass of Mars Colonizers would gather in Earth orbit.

Musk called them "a fleet." He noted that "if you want to have a colony, you've got to send a lot of ships at once. There's an optimal point every two years to send them, and you'd want the whole fleet to depart within a day or two."

The first trip would involve only one or two spaceships, he said. "Eventually you'd have hundreds or thousands of ships. If you want to get to a colony with millions of people, then you have to do something like that. I am saying that eighty thousand people will go at once every two years."

To Musk, the parallels to the British colonization of the New World remain striking. "It's just like America," Musk said. "How many English ships went to America the first time? One. And then if you fast-forward two hundred years, how many ships went from England to America? Thousands. So it would be something similar. There was hope in the New World. It may as well have been Mars."

Musk believes that millions of people may eventually want to go to Mars, and sign-ups for projects like Mars One indicate he may actually be correct. But he does not plan to play the role of Pied Piper, either. "It's not what do I want to do, but what are *people* going to want to do?" Musk said. "I don't know what people will want to do or what state the world will be in or where SpaceX will be at that time." But he added that "the system we're designing will be capable" of getting people there if they want to go. "I think hopefully there are tens of thousands of people going per orbital rendezvous by 2050," he said.

But let's back up a bit. Before those pioneers set off for Mars, someone will have to be the *first* explorer.

According to various other Mars missions (not planned by
Musk), before anyone can land on Mars and expect to stay a
while, two things must happen: an appropriate landing/living
site must be scouted and a lot of supplies must be sent from
Earth in advance. In an ideal scenario, the supply missions that
precede manned missions will robotically erect and maintain
habitats.

The Mars One mission has proposed such a system, using a
rover to construct a habitat from parts brought by cargo supply
missions that would precede astronauts. The technicalities
involved in landing cargo ships just so, then arranging for me-
chanics or robotics to connect them, then moving cargo around
and reconfiguring the capsules and attached inflatables are
legion, but certainly not unreasonable. However, to do all of this
by 2025, as Mars One has proposed, seems unlikely at best.

The Mars One plan appears to rely on using SpaceX *Crew
Dragon* capsules, which are projected to begin shuttling astronauts to the space station in 2017. The mission also appears to
rely on SpaceX's Falcon Heavy rocket. In development for several years, the Falcon Heavy is a combination vehicle drawing
heavily on the design of SpaceX's default rocket, the Falcon 9,
with two first-stage Falcon 9 rockets strapped to it. It is designed
to be able to lift four times as much payload to orbit as the
Falcon 9—it would have 4.5 million pounds of thrust, making it
the most powerful rocket on Earth (though it would have only
half the lifting capacity of von Braun's Saturn V rocket, which
launched *Apollo* spacecraft to the moon). SpaceX has been
signing up clients for Falcon Heavy rocket launches for several

years, but the test flights have been delayed. A NASA proposal created in 2011 dubbed Red Dragon called for using a Falcon Heavy rocket and a *Dragon* spacecraft for a low-cost Mars drilling expedition, but the mission was never finalized.

The Mars One timeline, published on its Web site, calls for sending cargo flights to the Red Planet in 2022, then sending four humans every two years after that, beginning in 2024. The Web site's home page currently shows six *Dragon*-like capsules neatly lined up on the Martian surface connected to one another by tubes. That's basically the strategy that many Mars enthusiasts, like Robert Zubrin, founder of the Mars Society, have been backing for years. The plan is heavily dependent on significant cooperation from SpaceX. Mars One says on its Web site that it has "visited" SpaceX and received a letter of interest. But no deal has been inked between Mars One and SpaceX, and Musk is dubious that the Falcon Heavy rocket can be used for a Mars trip. His Mars Colonizer rocket "will have three times the thrust of the Falcon Heavy and twice the thrust of the Saturn V." Meanwhile, there are likely to be many higher-priority customers for both *Dragon* spacecraft and Falcon Heavy rockets that could put Mars One farther behind. By the middle of 2014, Mars One had raised about $600,000 from donations. That is less than 1 percent of the cost of launching a *Dragon* spacecraft into low Earth orbit on a regular Falcon 9. Mars One charges applicants for astronaut positions a fee, which will bring them a few million dollars. It is also pursuing broadcasting rights, in the reasonable expectation that a Mars trip could be the most popular reality TV show of all time. Regardless, Mars One has a long way to go to raise the $6 billion that its CEO, Bas Lansdorp, says will be needed just for the first crew.

So far, Mars One seems to be an optimistic group of people who want to settle Mars but haven't yet put their pocketbooks where their hearts are. Other organizations have similarly vague proposals.

By contrast, though he, too, has revealed little about any detailed plans, when Elon Musk says SpaceX will put humans on Mars, and when he envisions millions of people there—a more far-fetched version of life on the planet than most have suggested—it is not difficult to believe him. That's because he has done the impossible before. First, he completely revolution- ized the 110-year-old automobile industry by cofounding Tesla Motors. Many people laughed at Tesla's early days, insisting that a world of electric autos was at least fifty years into the future. Yet only two years after the Model S began to be sold, there are an estimated seventy thousand on the road, and a Tesla owner can drive his or her car from coast to coast or up and down each coast with ease, stopping for free "fuel" at the 174 (according to Tesla Motors) Tesla charging stations. If you put solar panels on your home, you can drive a Tesla on sun- shine. Malls and parking lots everywhere increasingly feature multiple charging stations for electric autos, many of them free. The popularity of Tesla cars shows no sign of waning, and Musk is gearing up to produce 500,000 cars per year by 2020. Automobile companies including Ford, Toyota, and General Motors are scrambling to catch up. Long before they do, Tesla will introduce an affordable electric auto for the mass market. Within a decade, the internal combustion engine automobile is likely to look exactly like what it is—a machine that converts gasoline into much more heat than forward motion, a bizarre antiquity. Now Musk is doing it all over again with SpaceX—revolutionizing the way we get into space.

Prodded by Musk's bold plans and NASA's commitment to eventually using its *Orion* system to land humans on Mars, every spacefaring nation has joined the race to establish a presence on the Red Planet. In 2016 the European Space Agency will partner with Roscosmos, the Russian Federal Space Agency, to launch a Mars orbiter. (It's worth noting that this is not the ESA's first Mars mission; it sent the *Mars Express* to the planet in 2003.) The orbiter will measure trace gases—those that comprise less than 1 percent of the atmosphere. In 2018, the two organizations plan to land a rover on the Martian surface. The Russians have also discussed building an enormous rocket to rival NASA's Space Launch System, which could conceivably take a manned mission to Mars around 2030. Meanwhile, the Chinese have announced plans to send a Mars rover similar to their lunar rover to the Red Planet around 2020.

Imagining life on Mars . . .

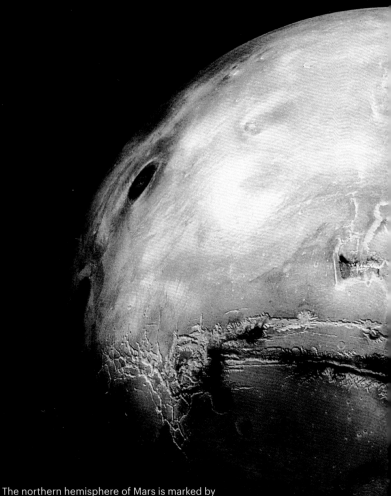

The northern hemisphere of Mars is marked by vast sandy plains, made up partly of red iron oxides. The jagged Valles Marineris canyon near the equator is nearly five miles deep and almost as wide as the United States.

ABOVE A 1954 *Collier's* magazine cover image sparked the public's growing interest in space travel.

OPPOSITE This 1954 depiction of the Mars landing, as imagined by a magazine illustrator, was based on the work of German SS officer turned engineer Wernher von Braun.

Bonestell and Miller

OPPOSITE The *Curiosity* rover snaps a composite selfie while drilling rock in a sandstone formation known as Windjana.

OVERLEAF Wheel tracks mark *Curiosity*'s path over a Martian dune.

OPPOSITE SpaceX designed the *Dragon* spacecraft for delivering crew and cargo to Earth orbit. CEO Elon Musk has a far larger and more complex spacecraft in mind for a Mars journey.

ABOVE The *Crew Dragon*, SpaceX's next-generation spacecraft, can carry seven astronauts. It is targeted to launch by 2017 under NASA's Commercial Crew program. Inspiration Mars, a non-profit founded by Dennis Tito, has proposed using the *Crew Dragon* spacecraft for a 580-day husband and wife flyby of Mars in 2021.

ABOVE The SpaceX Falcon 9 rocket (pictured) and *Dragon* spacecraft completed six cargo trips to and from the International Space Station within the past three years. Currently in development, the Falcon Heavy rocket will be the most powerful rocket in operation and will be capable of launching manned flights to the moon and even Mars.

OPPOSITE When a Falcon 9 rocket takes off, its first stage operates on nine Merlin engines designed by SpaceX. The rocket can lose up to two engines and still complete its mission.

OVERLEAF When the SpaceX *Dragon* spacecraft first berthed with the International Space Station in 2012, it proved that private companies could accomplish advanced feats of space travel that only governments had been able to do before.

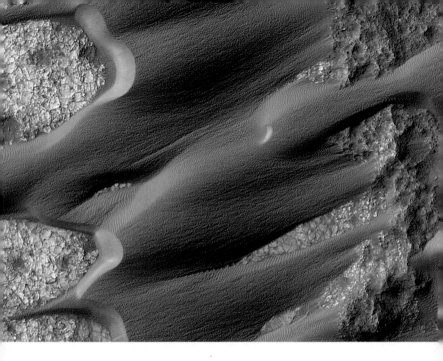

ABOVE NASA tracks seasonal and annual variations in the Martian winds by watching active dune fields like this one. The dunes' peaks can be up to a kilometer apart.

OPPOSITE Wind shapes the dunes in this crater into a V-like formation, akin to migrating birds in flight.

OVERLEAF This rugged landscape in the magical-sounding Noctis Labyrinthus region of Mars shows a bright network of ridges juxtaposed with dark sand dunes. The dunes, which migrate across the planet's surface with the wind, get their dark hue from iron-rich materials in volcanic rocks. Earth's pale dunes are predominantly quartz.

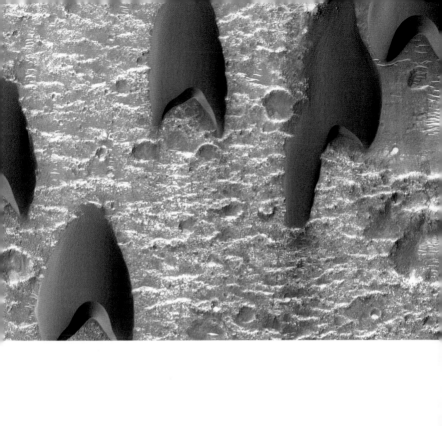

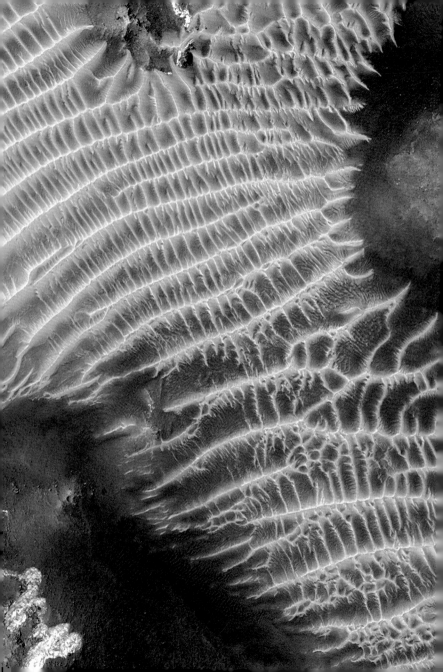

OPPOSITE Water ice makes up most of Mars's northern polar ice cap, providing clear evidence that the planet does possess the (frozen) liquid of life.

OVERLEAF Water ice forms a pool in a thirty-five-kilometer crater near Mars's north pole.

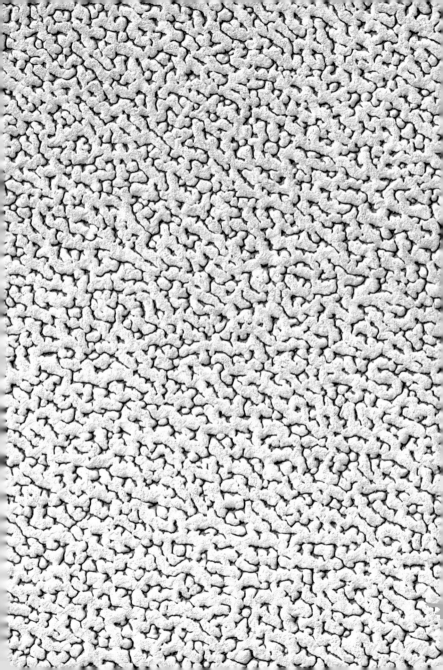

ABOVE The *Curiosity* rover located this seven-foot-wide iron meteorite dubbed "Lebanon" on May 25, 2014.

OPPOSITE The tailings around this sampling hole look like cat litter, and for good reason. They contain smectite clay, a key litter ingredient. Smectite-rich soil readily absorbs water and can support plant life.

OPPOSITE The frosty gullies on Mars, dusted mainly with frozen carbon dioxide (dry ice), but also a touch of frozen H_2O, are further evidence that the planet has water.

ABOVE Sedimentary rock strata frame the planet's Mount Sharp and represent millions of years of geologic history.

OVERLEAF A little speck of white light in the Martian twilight, just left of center, is our Earth.

ABOVE Light and shadow accentuate the outline of a draa, the largest class of Mars's sandy, wind-forged landforms. This draa has a wavelength of more than half a mile and probably formed over thousands of years or more.

6 Living on Mars

Humans need four things to survive on Earth—food, water, shelter, and clothing. Humans need five things to survive on Mars—food, water, shelter, clothing, and oxygen. The successful procurement of these five essential resources will secure humanity's future as an interplanetary species.

The Water Dilemma

We sustain brain damage after four minutes without oxygen, and the death threshold of oxygen deprivation is thought to be fifteen minutes. But no one is expecting us to find any oxygen on Mars. We will have to make our own. And we can do this with water—if we can find it. If we find water on Mars, we can make oxygen in several ways, including through simple electrolysis—passing an electric current through water. Thus, water is the most important element for human survival on Mars, especially because it's too heavy to bring all the way from Earth. If Mars does not have the water we think it does, we will not be able to live there.

Many years ago, when the various Mars orbiters and landers were but drawings on paper, NASA made an important decision—to "follow the water." The goal wasn't to focus on colonizing a planet; it was intended to help in the hunt for alien life. No water, no life. It now seems a bit ironic that NASA's insistence on investigating whether there is life on Mars has in fact led us to a completely different understanding: that there *can* be life on Mars—human life.

Intelligence gathered by various craft, including *Curiosity*, the *Mars Reconnaissance Orbiter, Mars Odyssey, Mars Express,* and even the *Viking* landers that date back to the 1970s, has established that there is, in fact, water on Mars. But it wasn't until the *Phoenix* lander touched down on the northern polar ice cap in 2008 that any probe could actually confirm without a doubt that frozen water exists on Mars, and that it is readily found in the Martian soil, called regolith.

Although Mars has a surface area only about 28 percent the size of Earth's, the amount of dry land is almost the same, because 70 percent of Earth's surface is covered by oceans, lakes, and rivers. None of Mars is covered by water, with one very important distinction. There may be more than a million cubic miles of water on the surface of the planet, but almost all of it is ice. Therefore, water may appear on Mars from time to time under special atmospheric circumstances, but until the atmosphere becomes much denser and surface temperatures become much warmer, water will rarely flow.

Much of the frozen water is at the north and south poles of Mars, some of it buried beneath frozen carbon dioxide. If all of it melted, the planet might be blanketed by an ocean hundreds of meters deep. That's a lot of water, but it is not nearly as much as geologic studies indicate once flowed on the planet. There are tens of thousands of river valleys on Mars, and numerous large lake beds. As much as a third of Mars may once have been covered by an ocean. A portion of the Elysium Planitia, a vast plain near the equator, may be a sea of fragmented ice as large as Earth's North Sea.

Ice on Mars seems to be abundant, but estimates of the percentage of water ice that can be found in a cubic foot of regolith varies widely, from about 1 percent to 60 percent. There are large numbers of mini-lakes of ice right below the surface of Mars, and many of them appear to be in a band close to the equator. Ponds of frozen water would be a very fortunate discovery for early settlers.

Some of the water that once flowed freely on the planet likely evaporated into space as Mars lost its atmosphere. The *MAVEN* spacecraft, now in orbit around Mars, is telling us a lot about that. Much of Mars's water may have receded underground, but most of it is probably bound up in surface ice. If Martian immigrants measure their early "wealth" in terms of water resources, they may be rich indeed. Had Mars proved to be as arid and waterless as it has sometimes seemed from telescopic images and early flybys, we might be looking at a far stranger planet to migrate to—Venus.

Finding water on Mars has not proven to be difficult thus far, but making it liquid will be a big challenge for the early settlers. A large part of the water problem will be the amount of human energy necessary to get it in the first place. Most of the water we find is likely to be ice mixed with regolith. That would make it a type of permafrost that will prove impenetrable without a jackhammer. Even then, liquefying the water may require quarrying techniques and lots of energy-expensive machinery, so the first colonists would be very lucky to find a small lake of pure ice.

The best of all possible scenarios would be finding liquid water. That could be possible underground. There is a great deal of speculation about water deposits under the Martian surface, but no one knows the reality yet. The first astronauts will need to be able to drill to at least moderate depths in hopes of finding a supply of liquid water. Extracting water from the surface of Mars, or from a well, is not rocket science, but it will involve tools that are specific to the job, including ovens and distillation devices (otherwise, drilling will create ice volcanoes, which will form as the liquid water freezes immediately upon surfacing).

One scenario is that chunks of regolith will have to be hammered out of the surface by astronauts, though in later supply missions the delivery of small bulldozer/truck vehicles will increase the amount of work each colonist can do. The regolith will be placed in ovens and heated until the ice evaporates as steam, which will then be distilled and filtered into drinkable water. There will be a lot of waste to deal with, and the process will demand energy—some of it available from solar cells, but much of it likely to necessitate a small nuclear reactor.

• • •

Very little that pertains to living on Mars in the early years will involve off-the-shelf equipment and supplies from Earth. Like Elon Musk's Tesla autos, almost every tool or device in use on Mars will need to have been carefully thought out. Someone drilling for water cannot discover halfway through the process that they have failed to anticipate a specific problem—a mineral deposit that requires a special drill bit, for instance. For survival to be a reasonable expectation, every circumstance must be anticipated.

So what might we do if the first astronauts on Mars find that all attempts to process regolith or drill for water or carve chunks of ice out of the Martian surface fail miserably? There's a good backup plan. We learned from NASA's *Viking*, the first spacecraft to land safely on the planet in 1976, that although the atmosphere on Mars may be thin, it is wet, in the sense that the level of humidity is often 100 percent. A research study published by the University of Washington in 1998 presents a device, called the Water Vapor Adsorption Reactor (WAVAR), that could extract enough H_2O from the Martian atmosphere to support human life. The paper notes, in part, that "the atmosphere of Mars is the most highly characterized and global water source on the planet.... [Although Mars's is] an extremely dry atmosphere compared to Earth's ... on the average, the atmosphere of Mars is holding as much water as it can on a daily basis, i.e., 100 percent relative humidity at night ... at most seasons and latitudes."

WAVAR uses water-adsorbing minerals called zeolites, which occur naturally as minerals on Earth and are easily made commercially. (They are used in industrial dehumidifiers to suck up water vapor from the atmosphere.) The WAVAR paper goes on to show how simple the process can be: "Martian atmosphere is drawn into the system through a dust filter by the fan. The filtered gas passes through the adsorbent bed, where the water vapor is removed from the flow. Once the bed has reached saturation, the water vapor is desorbed from the bed, condensed, and piped to storage. The design has only seven components: a filter, an adsorption bed, a fan, a desorption unit, a bed rotating mechanism, a condenser, and an active-control system."

To keep the mission's footprint and mass as small as possible, the WAVAR is designed to arrive on the Martian surface in a cargo shipment and begin making water two years before a crew of astronauts arrives.

To restate what may now be obvious: If Mars has the water that we think it has, continuous human life will be plausible.

The Oxygen Dilemma

Then there is the oxygen problem. When you run out of oxygen in a space suit, you can only breathe the carbon dioxide that you exhale for so long before you lose consciousness. Death is not far behind. Humans can tolerate no more than 5 percent of the air they breathe to be CO_2 for more than a very short period, partly because we have evolved to pass out from too much CO_2 as a defensive mechanism.

On this basis, Mars seems like a very hostile place; it has almost no oxygen in its atmosphere. The "air" on Mars is, according to readings from the *Curiosity* rover taken in 2012, 2 percent nitrogen, 2 percent argon, 95 percent carbon dioxide, and trace amounts of carbon monoxide and oxygen. The numbers change slightly as seasons change because during winter months some of the gases freeze at the poles and then are released in the spring. Although there's less than 1 percent free oxygen in the Martian atmosphere, there is plenty of oxygen on Mars. The secret is in the makeup of carbon dioxide, which is, by molecular weight, 28 percent carbon and 72 percent oxygen. If the Martian atmosphere is 95 percent CO_2, then at least 70 percent of the air on Mars, by mass, is oxygen. And even though the Martian atmosphere has only 1 percent the density of Earth's, that's a lot of oxygen.

The water that pioneers will carve out of Mars has even more oxygen in it—about 89 percent of water's mass is oxygen. And earthlings have become quite adept at a simple technology called electrolysis that can be used to separate water molecules so that they release oxygen. The process requires placing two electrodes in a tank with water in it and switching on an electric current to run through the water. Voila! Oxygen can be collected at one end of the tank, near the anode, and hydrogen can be collected at the other end, by the cathode. Hydrogen can serve as an excellent fuel and power source. Practically every high school chemistry student in America performs a version of this electrolysis trick in laboratory exercises. And there's a bonus: the hydrogen and oxygen, once separated, make ideal rocket propellants. The only problem with electrolysis is one that will frustrate early Mars colonists: it requires a lot of electricity.

NASA, fortunately, has already tackled the oxygen problem. When it launches the successor to the *Curiosity* rover in 2020, it will carry a type of fuel cell that will turn Mars's atmospheric CO_2 into oxygen and carbon monoxide.

The device is called MOXIE, for Mars Oxygen In-Situ

Resources Utilization Experiment. It uses a process similar to electrolysis in water, only with high-temperature ceramics in air. "A voltage across the ceramic selectively separates the oxygen ions that have been catalytically produced at the surface," says Dr. Michael Hecht, principal investigator for the MOXIE instrument project and assistant director for research management at MIT's Haystack Observatory. NASA's goal with MOXIE is less to prove that we can make breathable oxygen as to show that we can create an oxidizer for rocket fuel. Oxygen weighs far more than rocket propellants like hydrogen or methane, so NASA is obsessed with the concept of making it on Mars for return trips to Earth. It will be far more efficient not to have to carry fuel all the way to Mars for return voyages.

The MOXIE module on the next Mars rover will produce only about fifteen liters of oxygen per hour at standard temperature and pressure. That doesn't seem like a lot, but human lungs consume only five to six milliliters of oxygen per minute. "Bottom line, MOXIE can produce enough oxygen continuously for a human to breathe if he or she isn't very active," says Hecht. If MOXIE works as expected, NASA plans to scale it up by a factor of one hundred, although that will require a nuclear reactor for power.

"MOXIE is intended to be a one one-hundredth scale model of the plant that would eventually support a human mission," Hecht says. "The idea is that we first establish a robotic station containing the nuclear reactor and the oxygen plant, then send humans twenty-six months later, after confirmation that the O_2 tank is full and the reactor is working."

On Earth, people breathe in an atmosphere that is about 78 percent nitrogen and 21 percent oxygen. Although humans can breathe many mixtures of gases, including helium and oxygen, they cannot breathe a mixture of 20 percent oxygen and 80 percent carbon dioxide. The gas we mix with oxygen needs to be one that is nonreactive, or inert, such as argon or helium. Nitrogen is not usually considered an inert gas, but the bond formed by two nitrogen atoms is so strong that it tends not to react with other atoms.

The Food Dilemma

One crucial requirement for human survival on Mars will be food. Agricultural science is a highly developed area of study in many parts of the world, including the United States. Thus, many PhD candidates have spent years trying to figure out exactly how we'll grow plants on Mars. (Colonists will be vegetarians whether they like it or not, since animals are radically less efficient to grow.) If the first pioneers land near the equator, the days will be warm enough to use inflatable greenhouses. They will have to be well insulated and use passive solar techniques, such as heat absorbing stones exposed to sunlight all day, as well as electric heating to compensate for the sharp drop in temperatures at night. They will also require higher densities of atmosphere than currently exist on Mars. A typical Martian day around the solstice comprises about twelve hours of daylight and twelve hours of darkness. Estimates of the pressures needed inside a Martian greenhouse vary, but botanists expect to be able to grow plants in environments that contain about one-tenth the atmospheric pressure of Earth. We know from experiments conducted on the International Space Station that plants will grow in zero gravity, but no one knows for certain what effect Mars's gravity—about 38 percent as strong as Earth's gravity—will have on plants.

We know enough about the Martian regolith to have confidence that a lot of it will make good soil, though that will be somewhat dependent on the exact spot from which the regolith is scooped. Samples surveyed by landers and analyses of meteorites that have come to Earth from Mars indicate that there is a type of clay on the surface known as smectite, which is common on Earth and is often used in cat litter. The clay absorbs water readily and could be good for growing plants. Nevertheless, Martian soils may be too acidic or too alkaline and could require remediation and the addition of nutrients like nitrogen. Hydroponics—growing plants in nutrient-rich water, without soil—would offer the highest confidence of successful crops, assuming water is readily available and can be kept liquid.

Angelo Vermeulen, the biologist and artist who lived for months in a simulated Martian environment, says, "Personally, I'm not convinced greenhouses will work. There is too little sunlight and too much radiation. They look nice on a postcard from Mars, but they're not practical." Instead, he envisions hydroponic "growth chambers" buried under mounds of soil to thwart solar radiation or placed underground in naturally occurring lava tubes. "Growing food on Mars is all about control," Vermeulen says. "You need to finely control the environment. With LED lighting you can control the frequency, spectrum, and intensity of light. With hydroponics, water and nutrients are controlled tightly, and you can be more certain your crop will be successful."

Although early settlers will have to temper the Martian atmosphere's high percentage of carbon dioxide in growth chambers and greenhouses, heavy doses of the gas might make plants grow faster and produce larger quantities. "You can tweak the CO_2 and see what works best," says Vermeulen. The total amount of sunlight on Mars is about 60 percent of what we experience on Earth. At noon on Mars, sunlight delivers about six hundred watts per square meter of energy for growing plants. On Earth, the figure is about one thousand watts per square meter. Sunlight on Mars is thus equivalent to sunlight on Earth when the sun begins setting in the afternoon and hovers about 35 degrees above the horizon. You can visualize the Martian sun as being more like the weak sunlight received in the winter months on Earth in cities like Milan, Chicago, Beijing, and Sapporo.

Crops on Mars will need to be as nutritious as possible while taking up very little space. Although beans are high in protein and fiber and may well make up part of a Martian diet, research on exactly what the crop mix should be is still unfinished. Mushrooms can be grown successfully on the compost made from leftover parts of plants that humans don't eat. If Vermeulen designed the menu, insects would also be on it: "Insects should be part of the diet for astronauts. Grasshoppers and crickets are crunchy and full of protein. I like dried mealworms, too. For one of my projects, we fry them and put them in salads."

Lettuce and other leafy plants will be a luxury, but an important one. "Lettuce is not ideal. Its nutrient value is small and its volume is rather big. But it has a good psychological effect on people—it's crisp and fresh," says Vermeulen.

The biologist is mystified that people still think astronauts eat food that comes in a tube: "Astronauts want comfort food. They want to eat together. On the International Space Station, they asked for a table that had been removed to be brought back again so they could eat together. They want remembrance of where they come from—a link to their culture and identity. Chinese and Russian astronauts want to eat foods that are different from what an American might like."

A recent fifty-day experiment, sponsored by the Dutch Ministry of Economic Affairs, in a greenhouse in Holland produced a lot of optimism about our ability to grow crops on Mars, although it did not control for reduced gravity or sunlight. NASA provided Dutch growers with soil from Hawaii and Arizona that it thinks mimics the regolith on Mars. About 4,200 plants were grown from seeds, and every seed planted in the simulated Martian soil germinated. Cress, tomatoes, rye, and carrots were among the species that seemed to thrive in the Martian-like soil, which held water well, as expected. Other experiments continue, including Canadian experiments on Devon Island and a Mars Society greenhouse in Utah.

No matter how successful we are at growing food on Mars, in the early days it will be only a small part of the diet. Most food will come from Earth. "I don't think you'll ever get to the point where one hundred percent of what you eat you will be grown," Vermeulen says. "Honestly, if we get to where we can grow ten percent of our food, that will be a good start." That's partly because growth chambers and the apparatus they require are costly in terms of mass and energy. When it comes to space travel and living on another planet, mass and energy rule.

The Shelter and Clothing Dilemma

In the same way that plants will need special sheltering on Mars in the early frontier days, humans also will require specialized clothing and shelter to survive the Martian environment.

Metal rocket ships and inflatable buildings are not permanent solutions to coping with Mars's harsh environment. There are two kinds of radiation to deal with—radiation from the sun and cosmic radiation. Solar radiation is what earthlings experience when we get a sunburn at the beach—energized particles from the sun that breach Earth's atmosphere. Cosmic rays come from mysterious sources outside our solar system and are of much higher energy, and are therefore far more dangerous. On Earth, cosmic radiation is greatly ameliorated by the thick atmosphere. It is not stopped by mere skin—it can easily penetrate extremely thick metals, and it can cause chaos with electronics. Cosmic radiation comes in a constant stream of dribs and drabs. It is difficult to protect against because its energy level is very high. People living at high altitudes in the Rocky Mountains and pilots who fly transoceanic routes absorb lots of cosmic radiation. There is no question that increased radiation exposure correlates to an increase in the risk of death from cancer, although that risk may be increased by only a small percentage. Over time, almost any radiation exposure is bad for human health.

Right now, NASA is considering increasing the amounts of radiation exposure it will tolerate for astronauts on long journeys such as that to Mars. The thin atmosphere on the Red Planet should prove reasonably impervious to solar radiation. However, a direct-hit solar flare, which is rare but always possible, would, of course, be damaging to the long-term health of humans. People on Mars will need shelter with as many feet of regolith or rock above their heads as possible. A solar storm directed Mars's way will require shelter in a deep cave or the like.

Robert Zubrin's Mars Direct proposal, which he has refined over several decades, calls for building structures with vaulted ceilings, similar to those the Romans perfected, using bricks that could be crafted on Mars from regolith. A series of vaulted buildings situated side-by-side could provide extraordinary shelter from both the cold of Mars and solar and cosmic radiation, especially if the buildings are covered by ten feet or so of regolith.

Proponents of living on the Red Planet also say that astronauts should be able to use commonly found materials on Mars to create plastics for use in construction, as well as iron, and perhaps even steel and copper. All of these schemes are reasonably well thought out, but they require extraordinary amounts of energy and lots of specialized equipment. Zubrin imagines small trucks equipped with bulldozer blades to push and haul the incredibly hard, frozen regolith around.

Good strategies for shelter will evolve with experience. **Throughout history, humans have adapted brilliantly to their** surroundings, using local materials to produce shelters that make sense for specific environments. That will happen on Mars, but the earliest settlers may have to make do in caves or fissures or lava tubes to secure reliable protection from radiation. Eventually, when Mars is terraformed to resemble Earth, radiation hazards will be mitigated by increased atmospheric density.

Clothing must also play a role in protecting colonists from radiation and cold. And there is a problem unique to Mars that only clothing can solve: the lack of atmospheric pressure. On Earth, we live under a very tall pile of atmosphere. Hold out your arm and try to imagine the miles of atmosphere above you that push down on each square inch of skin. That atmosphere, on average, weighs 14.7 pounds per square inch at sea level. Our bodies push out against that constant pressure. On Mars, where the atmospheric pressure is less than one one-hundredth that of Earth, no human could live long without a pressure suit to match the outward push of the body. Unlike with the water, oxygen, food, and even shelter problems, the only solution to the pressure problem is to wear a pressure suit at all times, unless we prefer to live in pressurized containers.

Dava Newman, the Apollo Professor of Astronautics at MIT, is working on unpressurized, flexible, lighter spacesuit designs for "planetary locomotion." She says that "physiologically, you only need to provide about a third of the atmospheric pressure found on Earth," or less than five pounds per square inch. Her spacesuits are more like wearable clothing than bulky capsules. In her second-skin BioSuit design, she uses polymers and shape-memory alloys to create protective clothing that is more elastic and less cumbersome than current pressurized suits.

To gain mobility, Newman prefers not to use any more radiation shielding in a suit than is necessary. "I don't want to put those layers in a suit, because real shielding is heavy and massive. Do we need radiation shielding? Absolutely—but we may only need a little in the suit," because astronauts will spend most of their time in a rover vehicle or in a shielded habitat. "By the time we send humans to Mars," Newman says, "we'll already know the radiation environment from all the rovers on Mars and orbiters that we've sent over the past decades."

All of these questions can be reduced to one ultimate challenge for humans on Mars: How can we live in such an inhospitable environment? The answer lies in the creation of warming techniques that will increase the density of the atmosphere—in short, reengineering the entire planet to become more Earthlike. That process is called terraforming, and it will take centuries to complete. But it can—and will—be done.

7 Making Mars in Earth's Image

We humans have proven ourselves to be incredibly adaptable to unusual living conditions, easily accommodating environments as hostile as Amazonian rain forests and the perpetual ice sheets of northern Greenland. Nevertheless, we will surely tire of rebreathing units, the constant need to monitor oxygen levels, and the intense cold of Mars. So we will naturally turn our attention to reengineering the Red Planet's atmosphere to be breathable, and to warming the surface temperature.

At this point, it is worth noting that many scientists who have studied the evolution of Mars and who have consumed the information from the spacecraft we've sent there since the 1960s believe the planet once had flowing streams, lakes, and at least one ocean, along with a humid atmosphere and, possibly, life.

Fortunately for humans, there is a relationship between water, atmospheric density, and warmth. Here is the simple overview: If the temperature of Mars can be raised, it will likely release gases that are now frozen—gases that will boost the atmosphere, resulting in more density, which will create a greenhouse effect. Temperatures will rise, causing surface ice to melt, especially near the equator. Water will flow. Liquid water (and an appropriate atmosphere) will allow settlers to grow plants outside of greenhouses. In turn, those plants will add to the oxygen content of the atmosphere. As on Earth, the keys to life and ecology are inextricably linked together.

The process by which we will accomplish this reengineering is called "terraforming," which means shaping the earth. A more accurate moniker would be "planetary engineering." NASA has called it "planetary ecosynthesis." Although the creation of the term terraforming is often attributed to various science fiction authors, the astronomer Carl Sagan published an article in the prestigious journal *Science* in 1961 in which he proposed terraforming Venus to become habitable for humans.

Terraforming will be incredibly expensive, and it may take a thousand years before humans can walk the surface of Mars in an environment not unlike what one finds along the west coast of Canada. But if we engineer even a few degrees of temperature rise in the right parts of Mars, it will make life there far more pleasant than it will be on that day in 2027 when the first astronauts arrive. Dramatic changes in outdoor lifestyles can be accomplished within just a few centuries.

There are several scenarios for warming Mars, which is the first step in terraforming. In many ways, the sexiest approach, with the quickest results, is to simply build huge mirrors to reflect sunlight back onto the surface. It would be especially effective to reflect sunlight toward the south polar regions, where a thick layer of dry ice overlies water ice. Mirrors, however, would be the most expensive method of warming Mars, and technologically the most challenging. But if employed, mirrors could result in liquid water running in streams (during daylight hours, near the equator) within a matter of years. The mirrors used for this purpose would be flexible, like solar sails, and made of poly-amide films coated with an extremely thin layer of aluminum. And they would have to be unbelievably large—150 miles across. They would likely be too heavy to launch from Earth, so would have to be manufactured on Mars. It's possible that we could recycle a solar sail used for a cargo resupply spaceship built on Earth that takes the long way to Mars. The sail would provide part of the propulsion necessary for the trip, and, upon entering Mars orbit, it could be removed and ferried to a location where it would reflect sunlight back onto Mars. Mirrors could also be remarkably low tech; they could be placed at specific locations where the sun's rays would constantly push them away from Mars, while the gravity of the planet exerted an opposite and equal pull on them. They would become a type of satellite called a statite.

Robert Zubrin favors this warming scenario, and calculates that a single mirror 150 miles across could warm the south polar region of Mars by 18 degrees Fahrenheit. This would be enough of a rise in temperature to release vast quantities of carbon dioxide, a powerful greenhouse gas, into the atmosphere. The CO_2 release would cause something of a runaway greenhouse effect, melting surface water ice in the regolith that would in turn release water vapor, another powerful greenhouse gas, into the atmosphere. A mirror three hundred miles across might double the increase in temperature.

Another warming scenario within the realm of plausibility involves going into the asteroid belt and locating a large chunk of rock containing frozen ammonia. Eventually, for humans to breathe on Mars without specialized equipment, we will need a buffer gas in the atmosphere. On Earth, that gas is nitrogen, which makes up about 78 percent of the air we breathe. Ammonia (NH_3) is composed of nitrogen and hydrogen. If an asteroid with a lot of ammonia could be steered into a collision course with Mars, the resulting impact would do at least two things: create heat to help warm Mars and raise the level of greenhouse gases. The impact alone of a large asteroid on the planet's surface could raise the temperature of Mars by 5 degrees Fahrenheit. Unfortunately, there could also be disastrous results: an asteroid colliding with Mars could create a nuclear winter scenario, throwing up so much debris into the atmosphere that the planet would actually cool before it warms, thus greatly postponing the time frame for terraforming. Furthermore, ammonia is caustic, and a large amount of it in the atmosphere could create worse conditions for humans than would more carbon dioxide. Ultimately, though, the sun's rays should break down the ammonia into hydrogen and nitrogen. Part of the hydrogen would react with the iron oxide in the regolith and produce water. Some of the hydrogen would also likely dissipate into space, because gravity on Mars is weak.

(A largely impractical solution for warming Mars involves sending robotic spacecraft to a place like Titan, a moon of Saturn rich in hydrocarbons, and finding a way to vacuum up liquid methane, which flows in streams and makes up small oceans on the moon's surface, and transport it to Mars. If hydrocarbons such as methane were released into the Martian atmosphere, they would produce the greenhouse gases water vapor and CO_2.)

We have learned the hard way on Earth that certain fluorine-based gases are much more potent as greenhouse gases than CO_2 or water vapor. Chlorofluorocarbons, or CFCs, are an example. On Earth, they are powerful greenhouse gases that have been banned from spray cans and refrigerators and air conditioners worldwide because they destroy the ozone layer. But on Mars, they might be just the solution. The elements that could be used to create perfluorocarbons (PFCs) in a factory on Earth are thought to occur naturally on Mars. For decades we built such factories here to manufacture the gases that allowed our refrigerators and air conditioners to function. We have that technology down pat. But to create enough PFCs to make a difference in the Martian atmosphere would require huge factories operated by thousands of people, a scenario not likely to come into play before Mars has its first city.

The cheapest way to warm Mars may be to use bacteria that convert nitrogen and water into ammonia or that can create methane out of water and carbon dioxide. The catch-22 here is water. We need to warm the planet in order to have liquid H_2O, but we can't warm the planet without liquid H_2O. This is a problem ripe for the likes of scientists such as J. Craig Venter, one of the first people to map the human genome. Venter has long been trying to modify existing microbes. For example, oil companies could use bioengineered bacteria in an old oil well that still contains up to 20 percent of its original oil but which can't be easily pumped out. The right bacteria might feast on the oil and release methane—natural gas—as a waste product.

We are on the cusp of creating new species of bacteria that

can be engineered to perform specific tasks. If new bacterial forms could be grown to live on mineral deposits in the Martian regolith and spit out PFCs, Mars would soon be a much warmer place. Even if we use existing bacteria to produce ammonia and methane, Mars would be a good deal warmer within a few decades. The methane and ammonia produced would also help shield humans from solar and cosmic radiation.

The problem with employing new species of bacteria is that they may not be easy to stop once they are started. In the 1930s, American farmers were given kudzu seeds to plant to stop soil erosion. Kudzu is not native to the United States and is classified as an invasive species. Now its vines strangle much of the American South.

All things considered, crashing asteroids onto Mars's surface or creating genetically modified bacteria that emit greenhouse gases would be problematic at best. The simplest and most elegant solution, at least to begin the process, is to use a solar sail to warm the polar region. The problem with employing a solar sail—one that reflects sunlight back onto the planet—is primarily one of cost, but this method does not require using technology we haven't invented yet.

Once we warm Mars enough to get water flowing, we should be able to transplant rugged Earth plants to Mars, where they would multiply readily in the carbon dioxide–rich atmosphere. As the plants spread, they will begin to produce significant quantities of oxygen. But oxygen is not a greenhouse gas, and will tend to cool Mars. Because of the already thin atmosphere, Mars's weak gravitational field, and the fact that any greenhouse gases we introduce on Mars will eventually break down, constant replenishment and engineering of the Martian atmosphere will be necessary. Just as we build plants to clean and filter our fresh water on Earth, inhabitants of Mars will have to build plants to keep their atmosphere dense and breathable.

The interactions among many of the processes we set in motion on Mars can be both beneficial and dangerously unpredictable. On an optimistic note, the more ice we can melt to create flowing water, the more bacteria can break down nitrates and release nitrogen into the atmosphere—and thus the more acceptable the atmosphere will become to plants, which will add more oxygen to the atmosphere. The processes are astoundingly synergistic.

Reawakening Ancient Life-forms

There are some wild cards involved in terraforming Mars, including the possibility of reawakening ancient life-forms. If we keep in mind that water once flowed on the planet and that large lakes, rivers, and an ocean existed, as did a dense atmosphere, it is difficult to imagine that some form of life did not exist. We have found no evidence whatsoever that life was once present on Mars, but the *Curiosity* rover has proven that the planet has the basic chemical building blocks of life. Because liquid water is the key ingredient for supporting life as we know it, it's reasonable to believe Mars was not always as lifeless as it now appears to be.

In fact, one theory of how life began on Earth directly involves Mars. In the early days of the solar system, when asteroids and comets were flying around in great numbers, large chunks of Mars were knocked into space. If a life-form existed within those chunks, it could have made the journey to Earth and found a new home on impact. We have evidence that microbes can withstand space travel. Water is believed to have flowed on Mars around the time that life began on Earth. If life formed on Mars, it likely did so before it happened on Earth. This means our planet could have been seeded with life from Mars.

But the opposite has also been theorized. In Earth's early history, chunks of our planet were being knocked away by asteroids, too. Our moon may have actually been formed from the collision of a large object with Earth. If we find Earth-like life on Mars, the synergy between the two planets, and whether one seeded the other, will make for an astounding puzzle. Even more important would be finding microbes on Mars that are still alive. This discovery would be an incredible boon to emigrants, because those life-forms would be uniquely adapted to the planet. If those microbes were revived in numbers by newly flowing water, we can only guess what benefit they might be to the atmosphere and to more advanced forms of plant life. Even if early exploration turns up no obvious signs of life on Mars, we may not be able to tell whether life exists until the rivers start flowing again. Only then will we learn what might be lurking in the regolith, under rocks, and perhaps in deep thermal vents or in subsurface aquifers heated by geothermal effects.

As Mars warms, early settlers might wake up one morning and notice something like moss growing beneath their feet. If there is life on Mars that can be revitalized by warming, it could accelerate the adaptability of the planet to humans. Of course, it could also be extremely toxic, penetrating even the best spacesuit and killing every human on the planet. But what we know about life on Earth suggests that the latter is an unlikely scenario.

Another wild card concerns the life we will bring to Mars and how it might adapt there. No matter how thoroughly we try to scrub our spacecraft before leaving Earth, it likely will be loaded with hitchhiking microbes. It is probably foolish to assume the rovers we have already landed on Mars were sterile, because we know the clean rooms they were assembled in were not as clean as we thought. One way or another, we will introduce life to the Martian landscape. And it will likely find a way to flourish, especially if we get the water flowing.

There are shorter-term issues to terraforming Mars, such as heating the planet, and much longer-term problems, such as how to convert a toxic atmosphere to one that is breathable by humans. This difficulty was explored in the previous chapter, but is worth revisiting in the context of terraforming because breathable air is by far the most challenging, time-consuming, and expensive problem facing human settlements on Mars. The people and groups that have promoted Mars as a new frontier can be justifiably optimistic about our ability to utilize the technologies necessary to warm the planet and get water flowing. Up to the point of providing a breathable atmosphere, terraforming Mars is more about how much capital we're willing to throw at it than time. The fastest and costliest techniques could bring about radical changes on Mars in as little as a few decades. But oxygenating the atmosphere? That could be a thousand-year-plus job.

There are two huge difficulties. The first is that the air humans breathe on Earth is about 21 percent oxygen and 78 percent nitrogen, and the mix is crucial. A few percentage points less oxygen and we turn blue; a few percentage points more and we damage our lungs. The nitrogen we breathe is a placeholder—it doesn't react with our lungs, and we exhale it back out after breathing it in. But, by volume, it is the vast majority of what we breathe. Inert gases like argon—or a nitrogen-argon mix—are probably our best hope. Therefore, we not only have to find enough oxygen to pump into Mars's atmosphere, which is 95 percent CO_2, but we also have to displace most of the existing CO_2 with an inert gas. And, to complicate things further, even if we are able to readjust the atmosphere on Mars, the planet will tend to cool off when we reduce the carbon dioxide content. An atmosphere of oxygen and nitrogen or another inert gas is not a greenhouse-gas environment. On Earth, large amounts of water vapor, among other factors, tend to keep the planet warm. For example, when we've warmed Mars sufficiently for the ice to melt, a great deal of water vapor should enter the atmosphere. It should rain and snow.

The strategies proposed by scientists and engineers to oxygenate Mars are far sketchier and less precise than their other terraforming proposals. Not all the technologies needed to make a breathable atmosphere exist. We can make good guesses about how to do it, but we cannot be certain our attempts will work the first time we try. And we must be very cautious in our approach, because if we get it wrong, we may not be able to undo what we have done.

Even the most optimistic scenarios for reengineering the Martian atmosphere are projected to take nine hundred years. Yet within that time frame, humans are likely to make astounding progress, and there is reason to assume we will be successful. It has only been slightly less than fifty years since *Apollo 11* landed on the moon. Two or three hundred years from now, with our general knowledge base doubling every few years, we will have far deeper insights into the problem. And there's a catalyst in the mix—our ability to reengineer genetics, especially plant genetics, is advancing with lightning speed. As much as genetic modification can be evil words on Earth, it could be the answer to creating the atmosphere we need to live on Mars.

Let's explore what we now know about changing Mars's atmosphere. As we warm the planet and water begins to flow, it will hydrate deposits of nitrates and release nitrogen, which is essential to plant life. The more plants we're able to grow on Mars, the more oxygen there will be. As water flows over many of the oxidizing chemicals in the regolith, it will break them down and release even more O_2. Huge quantities of oxygen are tied up in the red dust covering Mars, which is composed mostly of iron oxide.

Small nuclear-powered machines could roam the surface of Mars, scooping up dust, heating it, and releasing oxygen. (But it's a bit far-fetched to imagine using a million or so lawn-mower-like devices, which would consume vast amounts of energy.) A better idea might be to follow Robert Zubrin's theory to populate Mars with bacteria and primitive plants to start the oxygenating process, which would then allow more advanced plants that produce far more oxygen to gain a foothold.

Solar and cosmic radiation will be a concern for plants, but as we warm the planet and the atmosphere becomes denser—albeit with carbon dioxide—radiation damage will be greatly reduced. As noted in the previous chapter, although the massive amount of CO_2 on Mars is a huge disadvantage to humans, it can be a boon to plants. Plants consume CO_2 and expel oxygen. The late physicist Richard Feynman was fond of saying that trees aren't really land plants—they grow from the air. They rely mostly on sunlight and carbon dioxide to grow, although most also need water from the ground. Plants should thrive in the Martian CO_2 environment, and our genetic engineering abilities should allow us to re-create plants that grow far better and faster on Mars than anywhere else. In the end, genetics may be the key to breathable air. We cannot expect plants as we know them to do enough work—they must be radically changed so that they can thrive in an environment with more radiation, less atmospheric pressure, and less nitrogen than that of Earth.

Plants, of course, are only part of the solution. Because we are rapidly increasing our ability to genetically engineer bacteria and other small microbes, we may be able to produce new life-forms that consume the things we don't need on Mars, such as CO_2, and spit out what we do need—namely, oxygen and nitrogen.

The assumptions that all this will take a thousand years do not take into account foreseeable advances in science and technology. In September 2014, NASA's *MAVEN* probe entered

Martian orbit. It is designed to study Mars's upper atmosphere and ionosphere to find out just how much of the gas remaining on the Red Planet is being ripped away by solar wind. The whole project—which will last a year—will attempt to discover what happened to turn Mars, which we know was once wet and humid and quite warm, into the arid, frigid place it is now. *MAVEN* may teach us a lot.

What we know for certain is that our knowledge of Mars is growing exponentially. Our ability to engineer life-forms is growing rapidly. We're getting smarter faster. Think about what we knew of biology and chemistry three hundred years ago, in the early 1700s. Then imagine what we will know three hundred years from now, in the early 2300s. Most of what we know now will seem quaint.

Do We Change Mars or Change Humans?

We are getting good at gene editing—manipulating the genes inside cells, including removing some and adding others. We are becoming more skillful at using viruses to enter the nuclei of human cells and alter the genetic coding inside them. So far this process has been directed toward curing diseases. But before long—possibly within fifty years—we will be able to genetically modify humans. We already are doing so, in many unseen ways. And nature has already done it. As much as 8 percent of our genetic code comes from viruses that attacked our bodies over the eons of human history, made their way into our cells, and changed the DNA there to help themselves replicate. We are copying that natural process—using viruses to enter human cells and change them. A company called Celladon in San Diego is in phase-2 FDA trials with a process for modifying heart muscle cells in people whose hearts don't pump efficiently enough. They are reprogramming the heart cells. And therein is an idea as big as Mars: Why can't we reengineer human lungs or human blood cells to split the carbon atom from the CO_2 molecule? It seems naïve to think we won't be able to do that within three hundred years.

Thus, the true answer to how we'll survive on Mars may not lie in how we'll change Mars, but in how we'll change humans. As frightening as that may seem, it is already within our grasp. We embrace it wholeheartedly when it comes to curing diseases or making us more resistant to them. We are rapidly reaching a point where humans, not nature, will be in control of our own evolution. There is no reason not to adapt that knowledge to make our backup planet a desirable place to live.

"I think astronauts will be augmented through gene therapy," says Angelo Vermeulen. "The human body is not meant for space travel. We know that some people are less affected by radiation than others. We will figure out why, and modify our genes so we can adapt."

Perhaps we won't be able to change humans within one lifetime to be able to breathe an atmosphere of carbon dioxide, but I expect we'll be able to genetically change human egg and sperm cells to effect that change in our progeny. Genetic engineering is not a fantasy. It is coming. Meanwhile, as time goes by and our terraforming techniques improve, we may make parallel advances in our ability to manipulate human genetics so that we arrive at a point when the atmosphere of Mars is 40 percent carbon dioxide and we have reengineered humans to be able to breathe air that is 40 percent CO_2. Genetics and terraforming may achieve a happy equilibrium.

Altering humans will seem to some more fantastic than altering a planet, but we are far more capable of the former than the latter at this point. It may be unnerving that we can use some of the powers we have always attributed to gods, but the genie is out of the bottle. We must embrace those powers to survive.

8 The Next Gold Rush

Unfortunately, the ultimate reason humans will try to change Mars into a planet they can easily live on without pressure suits and oxygen masks has nothing to do with the fact that we are destroying our home planet or the knowledge that we will need to become a spacefaring species before a dying sun either swallows the Earth or throws it out of orbit. People will go to Mars for the same reason the Spaniards went to the New World and farmers went to Sutter's Mill, California—to get rich. As with all former new frontiers, progress will be driven by the lure of starting over and finding wealth. Some of those who get rich off this frontier will do so by simply helping other people get there. Elon Musk clearly sees SpaceX as carrying on that tradition. He has already worked out the price of a one-way ticket to Mars.

After the first and second and third major waves of settlers explore Mars only to discover that gold is not lying in former streambeds, they may begin to focus on a statement buried on a NASA Web site about near-Earth asteroids: "The mineral wealth resident in the belt of asteroids between the orbits of Mars and Jupiter would be equivalent to about 100 billion dollars for every person on Earth today." The asteroid belt between Mars and Jupiter is extraordinarily rich in metals, but those metals are extremely hard to mine from Earth, partially because it's so expensive to overcome our gravity with rocket launches. But the gravity on Mars is weak, so rocketing to asteroids from there could be relatively inexpensive. And there's a bonus—the distance to the asteroids from Mars is far less than from Earth. Once settlement of Mars has been established, mining asteroids from Mars will be far cheaper and easier than using Earth as a base.

Musk, however, thinks mining asteroids from Mars will be too expensive if the metals are to be returned to Earth, and that simple commerce on the Red Planet will sustain a population. "The economic base of a Mars colony will be what people do on Earth—everything from opening an iron foundry to a Pizza Hut," he says. "In terms of what they transport back to Earth, I think it would be primarily intellectual property. So it would be entertainment, software, or something that can be transported with photons as opposed to atoms. Anything transported with atoms would have to be incredibly valuable on a weight basis, because the cost of transport back to Earth would be quite high. . . . In the architecture I have in mind, the return cargo [in spaceships from Mars] is less than the outbound cargo. Because in the return journey from Mars, you've got just your spaceship—no booster."

Meanwhile, we may need to mine those asteroids far sooner than anyone is guessing. As Earth moves toward a population exceeding eight billion people, it is running out of important metals—even basic metals we take for granted, such as copper. Many of the metals found in Earth's crust will likely become exhausted soon. Almost all of the easily mined gold, silver, copper, tin, zinc, antimony, and phosphorus we can mine on Earth may be gone within one hundred years. It is somewhat ironic to note that the metals and minerals most critical for manufacturing and the production of electronics actually came to Earth from asteroids. The original nickel, palladium, molybdenum, cobalt, rhodium, and osmium on Earth sank to the center of the planet when it first formed as a hot, molten ball. Those elements were pulled toward the Earth's core by its strong gravitational pull. As the Earth began to cool and the crust formed, circumstances in the formation of our solar system created a rain of asteroids that brought to Earth many of the rare and semi-rare metals we now mine for modern manufacturing.

NASA and many private space opportunists have already foreseen the market for metals obtained from the asteroid belt. But not everyone has figured out that it makes much more sense to mine those metals from Mars. Both Mars and the miniplanet Ceres would make ideal bases from which to launch asteroid mining operations, with expendable cargo ships sent into low-cost Hohmann transfer orbits to end up months later on Earth (or on Mars itself, which will require its own materials to build and maintain settlements). It's not much of a stretch to imagine the continuous shuttling of mining ships to asteroids from Mars and the establishment of Martian factories that will transform rare metals and elements into exotic finished devices that are then sent to the home planet. Imagine looking at an iPhone 30 and seeing these words: Made on Mars.

Asteroids are money in the bank. A forty-foot-long S-type asteroid (more than 15 percent of asteroids are S-type) is likely to contain more than a million pounds of nickel, gold, platinum, rhodium, iron, and cobalt. This has not gone unnoticed. A company reorganized in 2012 and renamed Planetary Resources, Inc., was set up to mine asteroids. Investors include former Google CEO Eric Schmidt and Google cofounder Larry Page. Planetary Resources' lead was followed in 2013 by a firm called Deep Space Industries. Its website currently looks like a science fiction film setting, with illustrations of CubeSats, scouting vehicles, and huge mining spacecraft assembled in space and never intended to enter a planet's atmosphere. The chief scientist at Deep Space is John S. Lewis, who has taught at MIT and the University of Arizona. He is the author of *Mining the Sky: Untold Riches from the Asteroids, Comets, and Planets*. This may all seem a bit like science fiction, but it's quite serious. Deep Space has secured NASA contracts to consult on asteroid exploration and is designing small spacecraft to scout potential mining sites. It intends to start drilling on an actual asteroid by 2023. NASA itself is also planning to send a manned *Orion* capsule to an asteroid by then.

Once a Mars base is reasonably functional, people will flock there. A simple look at the extraordinary number of people on Earth who migrate from country to country each year indicates that a huge population on Earth wants to go where the future seems brighter. It's part of the human spirit.

For example, it is not widely understood how quickly the American colonies grew. In 1620, the *Mayflower* carried 102 passengers to Plymouth, Massachusetts. Within ten years Boston had been founded as a city, and by 1640 more than thirty thousand new colonists had arrived, most of them dispersing westward. Jamestown, the first permanent colony in America, started with 104 settlers in 1607, and only 35 remained by the time the first supply ship arrived the following year. But by 1622, shortly after the *Mayflower* landed, the population of Virginia had grown to 1,400. The Mars settlement may not grow that quickly, although the length of a sea voyage across the Atlantic in the 1600s is comparable to the time it will take people to get to Mars on a spacecraft, and the cost, in relative terms, is not that different.

Mars will become the new frontier, the new hope, and the new destiny for millions of earthlings who will do almost anything to seize the opportunities waiting on the Red Planet.

Any discussion of settling Mars needs to recognize the slippery slope between need and greed. Although there are no native populations on Mars to overwhelm, there could easily be an unrestrained rush for material resources, devastation of the environment, destruction of sites valuable to scientific inquiry, and even a temptation to re-create indentured servitude. The Outer Space Treaty of 1967 and others that followed have attempted to make territory outside of Earth common ground. But humans have proven that they need laws to govern their behavior and enforcers as well.

If we get it wrong, if we repeat the mistakes of our past, the consequences could be devastating. But if we get it right, the potential benefits to the future of humanity are astonishing.

9 The Final Frontier

A little less than five hundred years ago, Ferdinand Magellan took five small ships and headed west into seas and lands never before seen by Europeans. Although Magellan's mission was to find a new passage to Asia, the outcome of his voyage was far from secure. No one knew if a ship could sail from the Atlantic into the Pacific, despite earlier exploratory voyages by Columbus and others. The fleet carried supplies for a voyage of up to two years, but the circumnavigation took three years. All but one of the boats was lost or destroyed, many crew members died, and Magellan himself was killed by a hostile tribe in the Philippines. Survival was difficult and often relied on pure human ingenuity.

That voyage changed everything. It was the dawn of the Age of Discovery. As continents and civilizations began to connect across the oceans, the size of the Earth essentially doubled. New resources beyond imagining were suddenly within reach. People were no longer residents of a city or a small area—they were residents of an entire planet. Distances that had once seemed unimaginably great became smaller. Empires were created and destroyed. Old Worlds and New Worlds collided. Plants, people, diseases, and cultures crisscrossed the Earth. Maize came to Europe, and horses went to the Americas.

Some economies flourished; others were destroyed. And every person's view of what the world was expanded, contracted, and multiplied.

A voyage to Mars will make the Age of Discovery look like a minuscule event in human history. Our world will suddenly encompass an entire solar system instead of one planet. Our abilities to geoengineer something as large as a planet will flourish. Trade routes that would have seemed impossible to previous generations will be established. The Earth will gain metals it desperately needs and the technical knowledge to very possibly save its environment. Opportunity for a new life elsewhere will give hope to millions.

We must work desperately and devotedly to save our home planet—there is simply nothing else like it anywhere that we know of. A view of Earth from afar clarifies how delicate our world really is. That extremely thin, blue haze surrounding Earth? That's our entire breathable atmosphere. Most of our oxygen is contained in the first five thousand feet or so of sky. A completely different view of the Earth, seen from a distant vantage point, will perhaps inspire hundreds of thousands of people. A wider sensibility and appreciation of how all things are woven together into a finite ecology will emerge, and humans may gain a far more sophisticated understanding of the meaning of life. Voyaging to Mars may give us the insight to see our planet in true perspective. We must never let go of that vision.

But can't we do both? Can't we be a spacefaring society and also find a real balance with nature on Earth at the same time?

Can't we learn how to better serve Earth by experimenting with terraforming Mars? Can't we learn from past mistakes that destroyed and disrupted civilizations when colonists invaded? Can't our new Age of Discovery be hopeful, bringing out the best in the human spirit while ensuring the preservation of our species—its incredible cultural achievements—and projecting us far into the future?

ACKNOWLEDGMENTS

Many thanks to Chris Anderson for insisting I write this book when I would have rather spent more time flying quadcopters with him, to Michelle Quint for brilliant structural editing, to Alex Carp for relentlessly keeping my facts straight, to John House for finding amazing Mars mind-candy on obscure websites, to Juan Enriquez for never letting me forget what humans can accomplish, and to my darling wife, Chee Pearlman, for constantly supporting me in spite of her belief that people have better things to do with their time than go to Mars.

Stephen Petranek's career of more than forty years in the publishing world is marked by numerous prizes and awards for excellent writing on science, nature, technology, politics, economics, and more. He has been editor in chief of the world's largest science magazine, *Discover*; the editor of the *Washington Post*'s magazine; founding editor and editor in chief of *This Old House* magazine for Time Inc.; senior editor for science at *LIFE* magazine; and group editor in chief of Weider History Group's ten history magazines. His first TED Talk, *10 Ways the World Could End*, has been watched more than one million times. He is now the editor of *Breakthrough Technology Alert*, for which he finds the investment opportunities that create true value and move the human race forward.

Stephen Petranek's TED Talk, available for free at TED.com, is the companion to *How We'll Live on Mars*.

Courtesy of TED

RELATED TALKS

Brian Cox
Why We Need the Explorers
In tough economic times, our exploratory science programs—from space probes to the LHC—are first to suffer budget cuts. Brian Cox explains how curiosity-driven science pays for itself, powering innovation and a profound appreciation of our existence.

Burt Rutan
The Real Future of Space Exploration
In this passionate talk, legendary spacecraft designer Burt Rutan lambastes the US government-funded space program for stagnating and asks entrepreneurs to pick up where NASA has left off.

Elon Musk
The Mind Behind Tesla, SpaceX, SolarCity . . .
Entrepreneur Elon Musk is a man with many plans. The founder of PayPal, Tesla Motors, and SpaceX sits down with TED curator Chris Anderson to share details about his visionary projects, which include a mass-marketed electric car, a solar energy leasing company, and a fully reusable rocket.

Stephen Petranek
10 Ways the World Could End
How might the human race end? Stephen Petranek lays out ten terrible options and the science behind them. Will we be wiped out by an asteroid? Eco-collapse? How about a particle collider gone wild?

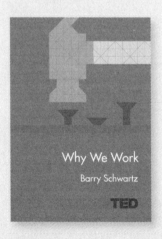

TED is a nonprofit devoted to spreading ideas, usually in the form of short, powerful talks (eighteen minutes or less). TED began in 1984 as a conference where Technology, Entertainment, and Design converged, and today covers almost all topics—from science to business to global issues—in more than one hundred languages. Meanwhile, independently run TEDx events help share ideas in communities around the world.

TED is a global community, welcoming people from every discipline and culture who seek a deeper understanding of the world. We believe passionately in the power of ideas to change attitudes, lives, and, ultimately, the world. On TED.com, we're building a clearinghouse of free knowledge from the world's most inspired thinkers—and a community of curious souls to engage with ideas and one another, both online and at TED and TEDx events around the world, all year long.

In fact, everything we do—from our TED Talks videos to the projects sparked by the TED Prize, from the global TEDx community to the TED-Ed lesson series—is driven by this goal: How can we best spread great ideas?

TED is owned by a nonprofit, nonpartisan foundation.